KRAKEN

THE

CURIOUS,

EXCITING,

AND

SLIGHTLY

DISTURBING

SCIENCE

OF

SQUID

KRAKEN

BY
WENDY WILLIAMS

ABRAMS PRESS
NEW YORK

For

BELLA JEAN THOMAS,

who was fed by the ocean as a child,
and who loved the ocean
more than
anyone else I know

In our hearts, we hope we never discover everything.

E. O. WILSON

CONTENTS

FROM VAMPIRE TO WALLFLOWER

All animals are the same but different.

—NEIL SHUBIN, PALEONTOLOGIST

I n the 1930s popular author and naturalist William Beebe cobbled together the world's first real-life deep-sea expedition with the help of fellow explorer Otis Barton. The team's exploration vehicle looked nothing like Jules Verne's sleek *Nautilus*. Small and round and crudely engineered by modern standards, the vessel was in diameter less than the height of a man, with three-inch-thick observation portholes and a bolted-shut door that imprisoned the men inside. The steel globe leaked, and to circulate oxygen internally, the men waved palm-leaf hand fans. Without an engine, Beebe's bathysphere dangled helplessly from the topside support ship like a ball of yarn suspended from knitting needles.

Clumsy and dangerous, it nevertheless did the job. Over successive dives, Beebe and Barton sank deeper and deeper, descending eventually 3,000 feet into a miraculous, twinkling, watery universe never before seen by anyone. To Beebe, the eerie life-forms pulsating with energy and light were ethereal. One deep-sea animal looked to him like "spun glass," another, like "lilies of the valley." On one dive Beebe narrated his descent to an ardent North American and European radio audience. Listeners hung on every word, as avidly as they would decades later when American astronauts walked on the moon.

William Beebe felt awe for most of earth's species, but for squid and octopuses he often expressed revulsion rather than reverence. He described a small Galápagos octopus as possessing a "bulging mass" of head and body with a "horrible absence of all other bodily parts which such an eyed creature should have,—nothing more than eight horrid, cup-covered, snaky tentacles, reaching out in front." The octopus's tentacles seemed to wave at him "as if in some sort of infernal adieu."

His description of the vampire squid was graphically lurid: "a very small but terrible octopus, black as night with ivory white jaws and blood red eyes," with "sinister arms" and webbing between the arms like a "living umbrella."

Let us try to forgive Beebe his prejudices. After all, his emotional responses reflected the spirit of the age. Long before Beebe's time another scientist—who apparently thought of this particular species as some kind of monstrous, diabolic chimera—had already named the animal: *Vampyroteuthis infernalis*, the Vampire Squid from Hell.

It's not hard to see why these men, denied the benefit of our modern scientific tools, found this particular little squid so repulsive. Half octopus and half squid (it may be an evolutionary stepping-stone between the two closely related groups), the foot-long vampire is semitranslucent with a jellyfish-like body texture. It has eight arms like an octopus, but it also has two bizarre antennae-like appendages that sometimes float in the water like cast-off fishing line. Probably evolved from feeding tentacles, these strange extremities seem to detect prey. *Vampyroteuthis's* usually blue eyes, the largest in the animal kingdom in proportion to body size, are capable of suddenly turning a devilish red, resembling the blazing coals of a Hadean fire. Hence, its hellish image.

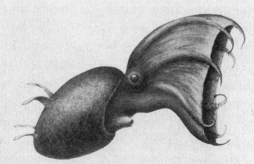

Vampyroteuthis infernalis

A vampire squid

Today we know that *Vampyroteuthis* was misunderstood. The vampire squid from hell actually lives a rather humble, mostly slow-motion existence thousands of feet below the sea surface, often floating peacefully in the water column. It may not do much of anything most of the time.

Tiny and not particularly powerful, it must sometimes even resort to self-mutilation as a defense. When threatened, little *Vampyroteuthis* bites off one of its eight arm tips, which are decorated with bioluminescent blue lights. As the severed arm floats away, its blue lights glow, luring the enemy in the wrong direction.

Recent undersea videos show us a vampire that's more like a wallflower or a shrinking violet than a demon from hell. Rather than fight, the beleaguered squid sometimes wraps itself up in its own arms so that it looks kind of like a deep-sea tumbleweed. It may cavort and tumble in the water until the confused predator gives up. If that doesn't work, the squid might distract its enemy by ejecting clouds of ink filled with glowing particles. Undersea videos show an animal that's often beautiful to watch. Were we to name this species today, we'd likely give it a kinder, friendlier name: maybe the wallflower squid, or the tumbleweed squid.

The more we get to know about the weird beings that live in the ocean, the less we fear them. There's very cool stuff in the deep sea, and some of that stuff, while worth knowing about in its own right, has also helped us live better lives. That's what *Kraken* is about. It's about how science and scientists work. It's about how we have learned that we are, more than Charles Darwin knew, truly kin to and beholden to all the other creatures of the earth. *Kraken* is the story of how the most serendipitous discoveries from the most unlikely creatures have revealed these basic connections, and about how field research, lab research, and ideas generated through scientific teamwork have not only provided insights into human biology but also created medical breakthroughs that have improved our lives. Over the past seven decades, Beebe's bathysphere has morphed into a myriad of manned and unmanned submersibles that have taken us all, as voyeurs if not as actual voyagers, into marvelous deep-sea universes. During those same years, we've also made remarkable journeys in genetic research and basic biology, aided, in part, by squid.

It turns out that the vampire squid is our distant cousin, albeit many (many) times removed, and that, curiously, we share a lot of basic biology. Moreover, tantalizing clues hint that some species of squid may be intelligent and capable of learning from experience. We've seen that the Humboldt squid, like dolphins, hunt in well-coordinated packs. Cuttlefish communicate with each other in intricate code, using a language of flashing colors and skin patterns. Octopuses build themselves houses. (They also like cast-off beer bottles, but prefer brown glass to clear, and short necks to long necks.) Some octopuses can untie silk surgical sutures.

· : · : · : · : ·

I came by my appreciation for cephalopods—squid, octopuses, cuttlefish, and the nautilus—quite by accident. On Cape Cod, where I live, squid are generally regarded as either restaurant food or fish bait. But when I spent a summer as a science journalism fellow at the Marine Biological Laboratory in Woods Hole, Massachusetts, several biologists told me that squid deserved the Nobel Prize for their contributions to human medicine. Bypass the scientists and go directly to the animal that's made the science possible, they told me. They were only half joking. That we can learn so much about our own bodies by studying such animals was, for me, a revelation.

Squid have even helped us understand the workings of our own brains. Without squid, neurosurgeons would be a little less well trained, obstetricians a little less well informed, and geriatricians much less knowledgeable about the aging process. In the near future, squid may help us cure Alzheimer's disease, improve camouflage for soldiers on the battlefield, and boost the health of babies born by cesarean section.

But cephalopods also deserve to be studied just because of their own uniqueness. "When you look into their eyes, you know there's something there," squid expert and Smithsonian scientist Clyde Roper told me.

I knew what he meant. When the animals stare so intently into our human eyes, they are seductive. With eight or more dangling arms and tentacles encircling their mouths, with the ability to change color and shape in milliseconds, with suckers as dexterous as our fingers and thumbs, and with eyes that are better than ours in some ways, they are enticingly, bewitchingly, exotically alien.

A WONDERFUL FISH

If you believe such things, there's a beast that does the bidding of
Davy Jones. A monstrous creature with giant tentacles that'll
suction your face clean off, and drag an entire ship down to the
crushing darkness. The Kraken . . .

—PIRATES OF THE CARIBBEAN

On October 26, 1873, Theophilus Piccot and an assistant known to history as Daniel Squires rowed out for herring over the icy-cold surface of Portugal Cove in Newfoundland's Conception Bay. Piccot knew the bay well. He had fished these waters hundreds of times. But on this trip, he and Squires saw something unusual floating in the distance below the Newfoundland cliffs.

It was quite large.

It looked, at least from a distance, something like an abandoned sail or debris from a wreck.

Hoping for valuable salvage, they rowed over. The two men found a quivering mass unlike anything they'd ever seen. They poked the mass with a gaff. It was a living creature. It reared its beak at them, which the men later said was "as big as a six-gallon keg." The animal's beak rammed the bottom of their skiff. From its head shot out "two huge livid arms." The animal then began to "twine" its arms around the boat.

The two feeding tentacles, several times the men's height and covered with serrated rings inside the suckers, shot out over the gunwales of the skiff, seeming to move with the speed of a lightning bolt. Fortunately, Piccot had a hatchet on board. He hacked away. He severed both tentacles, as thick as his muscular wrists, from the rest of the creature. The animal shot out gallons of ink which "darkened the water for two or three hundred yards." Then it sped away as the men watched. It was never seen again.

Piccot and Squires returned to port in St. John's, bringing with them what might well be one of the world's best-ever fish stories. They also brought back both severed organs, which had begun to stink almost unbearably. One they destroyed, not knowing its scientific value.

The other was saved by the local rector, who received the flesh as though he had received the stone with the Ten Commandments. Moses Harvey, like so many educated Victorians, was an amateur naturalist. He had followed the decades-long scientific controversy over the existence of a fabled sea monster. He may well have read, only months earlier, a paper by A. S. Packard published in *The American Naturalist* arguing for the existence of a very large animal, *Architeuthis,* in the North Atlantic. The animal had been given its scientific name years earlier by a Danish scientist, but there were still those who contested its existence.

Harvey understood the importance of having a genuine specimen. He had the 19-foot-long lump of flesh exhibited in the town's museum. He coiled the tentacle like a snake and had a drawing made. He also had a photograph taken. He sent a written report across the sea to the British *Annals and Magazine of Natural History.* The journal published the package under the title "Gigantic Cuttlefishes in Newfoundland."

The animal was, of course, not a cuttlefish (a small kind of cephalopod) but a huge squid. The misidentification is not surprising, given the mystery that then surrounded the species. Harvey's submission ended a scientific controversy that had existed for centuries and grown increasingly personal and even bitter as the nineteenth century progressed. Seafarers had long claimed that a massive, vicious animal lived in the deep sea. They said that the animal sometimes attacked ships and could tear a man to pieces. Whalers claimed the monster was as large as—if not larger than—a whale. They believed these monsters attacked whales. They had seen six-foot scars, made by what they thought were huge claws, on the skin of the sperm whales they took out of the sea. When they opened the whales, they found what looked like prodigious parrot beaks in the whales' stomachs. The whalers' stories were part of an eons-old tradition regarding an animal called by various names—"Kraken," "the Sea Monk," "the Great Sea Serpent," and even "the Great Calamary"—that lived in the sea. Odysseus's six-headed Scylla may have been part of that tradition, according to author Richard Ellis.

Reports of the animal had been sporadic and confused. The tales told by the frightened people who saw the animal were so varied that it was difficult to tell whether they were seeing the same species all over the world, or a wide variety of animals with only a few characteristics in common. Classical Greeks told of a hydra, a nine-headed serpent. The New England Pilgrims said they saw in the 1630s a "coiled sea serpent" on the rocks on the Cape Ann shoreline. In 1734, the Bishop of Greenland insisted he had seen a "web-footed serpent" during an Atlantic crossing. In 1851, Herman Melville's *Moby-Dick* described "the most wondrous phenomenon which the secret seas have hitherto revealed to mankind. A vast pulpy mass, furlongs in length and breadth, . . . curling and twisting like a nest of anacondas."

Throughout the nineteenth century, as people increasingly plied the sea, reports of "a wonderful fish" in the globe's oceans multiplied. Scientists remained skeptical. These confident— sometimes overconfident—men of the Victorian Age scoffed: How could the earth or the sea contain an animal so large that remained unknown to science? At that time, science theorized that no life could survive in the cold and lightless ocean depths, so the creature should have lived near the surface and been easily seen. No hard evidence existed to prove the sailors' claims. With little more than fishermen's tales, the scientists said, there was no reason to believe in the beast's existence.

Sailors and fishermen took umbrage. They knew very well that life existed deep down in the ocean. They had firsthand knowledge: Harpooned sperm whales often dove thousands of feet below to escape their fate and whalers routinely paid out thousands of feet of line to keep the animals from escaping. They also knew that the stomachs of these whales contained all kinds of unusual species that were rarely seen at the sea's surface. These strange beings had to live *somewhere* in the ocean.

Nevertheless, despite the specific knowledge provided by sailors and seafarers, science stuck to its dogma: Nothing could survive the water pressure deep below. No such thing as a giant squid could possibly exist.

In 1848 the matter came to a head. Peter McQuhae, captain of the British HMS *Daedalus*, reported seeing a 60-foot sea monster, nothing like a whale, floating on the water near the Cape of Good Hope. McQuhae wrote that he and his officers saw the thing at such a close range that, had it been a man, they would have seen his facial characteristics quite clearly. The animal moved at a speed of about 10 knots, the captain wrote.

Richard Owen, a paleontologist and a gifted scientific giant of his age who had coined the word "dinosaur," ridiculed the captain. Owen, not well known for his pleasing personality, may have felt some righteousness regarding the naming of cephalopod species, as he was the first scientist to describe the nautilus, the cephalopod that lives inside the beautiful pearly shell. Many sea peoples knew about the shell, which could float for hundreds and even thousands of miles once the animal inside was dead, but until Owen came along, no European scientist knew the detailed natural history of the animal that lived inside.

Owen was unwilling to believe in the existence of a humongous animal so closely related to the tiny nautilus. He did not just publicly disparage McQuhae's claim. He hacked away at the sea captain's personal credibility, implying in print that McQuhae was either a liar or a fool. According to Owen, the captain had seen nothing other than a very large seal or sea elephant (what we would today call an elephant seal).

McQuhae resented the implied slander, which subtly suggested that he wasn't equal to the task of ship's captain. McQuhae insisted that he certainly knew the difference between an elephant seal and a 60-foot sea monster. The battle raged on. Neither side would let the matter drop. Seafarers, scientists, and the British upper class continued to write treatises on McQuhae's sighting for decades after. As the years passed, more and more people came to accept that such an animal existed. "But science, incredulous, evidently will never be satisfied till it has a body to dissect," Sir William Howard Russell wrote, taking McQuhae's side in his 1860 book, *My Diary in India*.

Then came Theophilus Piccot's severed tentacle. Measured at 19 feet and tangible beyond dispute, the putrefying prize ended the argument. Piccot's animal came to be acknowledged as a squid—*Architeuthis*, the earth's largest then-known invertebrate. The word comes from the Greek, "archi" meaning "chief," and "teuthis" meaning "squid."

Among those vindicated was Jules Verne, whose 1870 smash hit *Twenty Thousand Leagues Under the Sea* tells of a gargantuan and malevolent monster that attacks a marvelous if seemingly fantastic electrically powered submarine (no such thing yet existed) and devours a crew member. Verne based his story on a similar giant squid sighting by a French sea captain, who had managed to bring home a tiny bit of flesh to prove his story. The French captain was also ridiculed by scientists, some of whom claimed the captain's prized flesh was probably little more than a decaying bit of plant life.

But by the 1880s, after the publication in a respectable scientific publication of Theophilus Piccot's excellent adventure, the controversy seemed settled. The existence of at least one species of giant squid seemed proven. In 1883, only a decade after Piccot's encounter with a live specimen, the International Fisheries Exhibition in London exhibited a massive model giant squid. *Sea Monsters Unmasked,* a pamphlet distributed at the exhibition that year by marine biologist Sir Henry Lee, suggested that many of the sea monsters written about over the millennia were nothing more than run-of-the-mill giant squid.

The public thrilled to the frightening confirmation of the existence of so awful an animal. The Fisheries Exhibition giant squid model was suspended rather ominously above the heads of long-skirted Victorian ladies and high-hatted Victorian men. The model squid was a slightly bug-eyed squid, with its two feeding tentacles stretched well beyond its other appendages. Designed to create awe in the public's mind, the model wasn't entirely anatomically correct, but it was fairly well done for an animal whose existence had been, only a decade earlier, very much in doubt.

A giant squid at the International Fisheries Exhibition (1883)

. : . . : . : .

Evolutionarily speaking, it took a long time for the giant squid to appear. Five hundred forty-two million years ago, about four billion years after earth came into being and perhaps three billion years or so after the simplest life-forms took shape, there occurred one of the most important events in the history of our solar system: the sudden radiation of life forms in earth's oceans.

This milestone, called the Cambrian Explosion, was a bit miraculous, a bit bizarre; extraordinary, but perhaps at the same time, scientifically speaking, inevitable. Before this divide, life existed on earth, but, quite frankly, it didn't amount to all that much, at least not to our modern eyes. There were no plants. For much of that time, there were no animals. From our point of view, it would have been a rather boring planet. But there was a lot going on behind the scenes. The stage was being set. Simple viruses and bacteria were probably around for quite a while, but evolution merely crept along. Then fungi and algae and simple single-celled animals proliferated.

Their presence freed up for the first time large amounts of oxygen in the atmosphere. Gradually, more complex animals evolved. But there was nothing of great size, nothing that would impress most of us today. Then, in a few tens of millions of years before the Cambrian, animals somewhat resembling a few of today's animals finally evolved.

About 555 million years ago, or 20 million years before the Cambrian Explosion, tiny *Kimberella* appeared. In some ways, the fossils of these tiny animals looked like jellyfish, and scien-

A Kimberella

tists at first assumed that's what they were, in part because
mainstream science held that sophisticated life probably did
not exist before the Cambrian. But as more examples turned
up, closer inspection revealed several mollusklike features.
The creature had a protective shell, a soft body, and probably
a radula, a tonguelike structure common to most mollusks even
today. Today, many scientists believe that *Kimberella,* only a few
inches long but apparently plentiful in the earth's shallow seas,
may be the earliest known ancestor of today's squid, including
the giant squid.

If it's true that *Kimberella* was actually a mollusk, paleo-
biologists will have to rethink the earth's evolutionary timeline.
Scientists postulate the existence of a proto-animal called
urbilateria from which much of the planet's animal life has
evolved. From this hypothetical "first animal" derived two
superphyla or major divisions—the *deuterostomes,* from
which we descend, and the *protostomes,* to which mollusks,
including the cephalopods, belong.

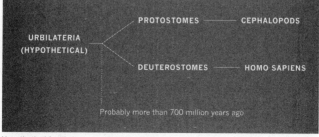

PROTOSTOMES ———— CEPHALOPODS

URBILATERIA
(HYPOTHETICAL)

DEUTEROSTOMES ———— HOMO SAPIENS

Probably more than 700 million years ago

Hypothetical family tree

Kimberella's pre-Cambrian appearance means that the hypothetical *urbilateria* and its two superphyla must be far older than scientists once believed, perhaps having evolved well before 700 million years ago. And because humans and squid share so much basic biology, like the camera eye and the neuron, *urbilateria* may also have possessed the foundation for some of this basic biology. In other words, the evidence suggests an exciting idea: that very early life on our planet may have been much more sophisticated than we currently believe. The more we know about cephalopods, the more progress we will make in unraveling this mystery.

. : . : . : . :

About 100,000 known species of mollusks live on our planet, although there may be another hundred thousand mollusk species not yet discovered and named. Mollusks live in every one of our ecosystems, except the desert, which is too dry for these moisture-loving animals. All mollusks are soft-bodied, like worms. But unlike worms, mollusks are not segmented.

Mollusks have a head, a main body, and a foot (even *Kimberella* appears to have been organized this way). They are often, but not always, protected by a covering shell. The mollusk's body contains vital organs like the stomach and intestines. The head contains sensory organs like eyes and either a simple nerve center or a true brain. The mollusk's foot is a tough muscle controlled by nerves connected to the animal's head. It's called a foot because the animal uses it as such, flexing its muscles to creep along the seafloor in search of food or to escape predators.

Mollusks also usually have a radula, a tonguelike, somewhat firm structure in the animal's mouth. Covering the radula are numerous rasping, tough, tiny hooks that remind me of Velcro teeth. In some mollusks, these hooklike teeth scrape algae and other food off objects like rocks or off the seabed itself. In other species, the radula is part of the digestive system and abrades prey into small, consumable bits; it is sometimes so effective that the swallowed food resembles pabulum.

If *Kimberella* was truly the first mollusk, it was an astonishingly successful organism: Today, roughly a quarter of the sea's animal species are probably mollusks. As befits one of the planet's oldest phyla, modern mollusks vary widely. On the one hand, some are small enough to live between grains of beach sand and weigh less than an ounce. On the other hand, the giant squid and the colossal squid, weighing hundreds of pounds, are also mollusks. In addition, the mollusk group includes scallops, mussels, abalones, and snails.

During the Cambrian Explosion, most of the planet's major animal groups, or phyla, appeared. The seas filled with life. Cambrian animals were not particularly large at first, but they were plentiful and innovative. Jaws appeared. Eyes appeared. Nature began experimenting with weaponry.

Early versions of claws appeared. Strange grasping appendages used for catching prey extended out from the bodies of some of the animals. The peacefulness of the early oceans disappeared in a maelstrom of predation. Thousands upon thousands of species killed or were killed, appeared and disappeared during this roughly 50-million-year period. In all this chaos, the *Mollusca* thrived. Early mollusks were tiny and were probably scavengers. But by the end of the Cambrian, the first cephalopods—hunters rather than scavengers—had evolved. Tiny *Plectronoceras* seems to have crawled snail-like along the seafloor, carrying an upright shell that looked something like a cow's horn. Inside that protective shell was the animal's main body, including vital organs.

Throughout the rest of the Cambrian, the cephalopods developed the body and lifestyle necessary to a formidable assailant. By the beginning of the next major evolutionary period, animals that we today would easily recognize as cephalopods were poised to rule the seas. Since then, for much of our planet's history, cephalopods have ranked high among the ocean's most dangerous, most prolific, and most triumphant predators, even, at times, holding the top-predator status of today's sharks.

Cephalopods are extremists, ranging in size from the giant and the colossal squids to the *Octopus wolfi,* which, at a bit more than half an inch in length and weighing in at only a fraction of an ounce, may be the world's smartest Lilliputian animal.

Cephalopods are also hardy. They are experts at adaptability. Teuthologist (cephalopod scientist) James Wood of the Aquarium of the Pacific calls them the "weeds" of the ocean. Since the Cambrian, there have been several major and many localized extinctions of animal life. Impressively, although some cephalopod species became extinct during some of these periods, cephalopods as a group not only survived but prospered.

Scientists can trace a long, steady line from *Plectronoceras,* the possible foundation animal for all cephalopods, to the roughly one thousand or so species of cephalopods that exist on the planet today—squid, octopuses, cuttlefish, and nautiluses. As befits a group of animals with such a long evolutionary history, cephalopods are quite varied. Some have elegantly specialized capabilities: The blanket octopus has developed a special immunity to the Portuguese man-of-war's poisonous tentacles and can rip them off with its own arms, then wave them in the water to warn off predators. The aptly named flamboyant cuttlefish is a species that has developed a poison so effective that it doesn't need to bother with swimming. Instead, it usually just lumbers across the seafloor using its arms as legs. Sometimes, when it raises bits of flesh on its back, it looks to me like an ancient armor-plated dinosaur. The colossal squid has soccer-ball-size eyes and unique swiveling hooks on the ends of its feeding tentacles and seems to hang in the midwater, waiting for a victim to swim by. In contrast, the Japanese flying squid weighs less than a pound, lives near the water's surface, and escapes its predators by jetting quickly through the water and into the air. The long-armed squid lives in the dark more than a mile below the surface and has ten super-long appendages that it may drag across the seafloor to sweep up bits of food.

As varied as they are, most cephalopods share a few basic characteristics. First and foremost, they have highly developed senses, like sight and scent, with which to respond to and adapt to ever-changing ocean conditions. (This is less true for the nautilus, an exotically beautiful cephalopod that never evolved a shell-free lifestyle and is not as intellectually advanced as the others.) Most cephalopods have a brain-to-body-weight ratio that places them above fish and reptiles and just below most birds and mammals. While the ratio of brain to body weight does not always correlate with intelligence and the ability to learn from experience, it is one factor scientists look at.

All cephalopods live in salt water; none have adapted to fresh water. Most live fast and die young, usually after they reproduce. A few small species of squid live for only a few months at most. At the other end of the spectrum, the nautilus may live for as long as fifteen years.

Most have the ability to change the color and texture of their skins, although refinements in this ability vary considerably from species to species. In general, squid are less talented at this than cuttlefish. In some octopuses, this ability is highly developed, but other octopuses have only a few colors in their repertoire.

All cephalopods are predators, and to improve their success at hunting, most modern cephalopods have traded the protection of shells for the convenience of high mobility. Without burdensome shells, they can swim through the sea like fish. Some squid species are so streamlined as to look like torpedoes. They can navigate the water in short spurts at speeds of up to twenty or twenty-five miles per hour, or as fast as some sharks.

A squid

All cephalopods have three basic parts. As in other mollusks, the muscular mantle contains vital organs like the stomach. The head contains the eyes and the buccal mass, or mouth area, including the beak. The arms or tentacles are not attached to the body, but encircle the mouth. To us, this seems backward: Even the word *cephalopod,* meaning "head-foot," alludes to this unusual arrangement.

Squid are called decapods, because they usually have ten appendages—eight arms and two much longer feeding tentacles, which are often carried tucked up close to the body. When the animal spots prey, the elastic tentacles shoot out. In the blink of an eye, the tentacle tips grasp the victim and then retract like rubber bands. The victim is injected with paralyzing toxin, shredded or sometimes liquefied, then eaten.

An octopus

Octopuses are called octopods because they have only eight arms. They lack feeding tentacles but many can instead envelop their prey in webbing that encircles their eight arms surrounding the beak. Cuttlefish also have eight arms, as well as two feeding tentacles that operate like the squid's feeding tentacles. The female nautilus has about fifty arms; the male, about ninety. Nautilus tentacles are individually much less powerful than those of the more modern cephalopods, but in aggregate the wriggling mass is more than strong enough to capture prey. At the tip of each arm are taste buds that explore the ocean and the seabed looking for available food.

Most cephalopods have suckers on their appendages. In some species, most notably octopuses, these suckers are extremely refined in their capabilities. These muscular suckers are controlled by individual nerves and can operate independently of each other. Some octopus suckers, like our own fingers and thumbs, are dexterous enough to manipulate objects. These octopuses can pass an object down an arm, grasping it firmly and rolling it along the arm from one sucker to the next. This reminds me of the way bodies are passed overhead from one person to the next in a mosh pit.

Most cephalopods have ink sacs that expel the ink through a funnel out into the water. The ink is used to help camouflage the animal, creating a smoke screen consisting either of a cloud of dark color in the water that masks the animal, or incredibly, a pseudomorph, a mucus-filled inky form which may actually take the shape of the animal itself. The predator may attack the pseudomorph instead of the real animal.

Intriguingly, cephalopod ink sometimes contains dopamine. In our own brains, dopamine is a neurotransmitter that produces euphoria. It's central to our reward system and involved in sex and drug addiction. The presence of the same molecular compound in squid ink is mysterious. Does a predator get high on the dopamine in the squid ink and give up its hunt? No one knows, but dopamine's presence in cephalopods implies that the molecule has been around, in one role or another, since the earliest days of evolution. If we ever come to understand its role in squid ink, perhaps we'll understand something more about our own predilections for addictive behavior.

Cephalopods also have a funnel, a muscular tube that's a kind of all-purpose tool, like an elephant's trunk. It acts as part of a bellowslike two-stroke system that jets the animal through the sea: The muscles in the mantle draw seawater inside the animal's body, then the funnel expels the water. Sometimes the flow through the funnel is powerful enough to allow the animal to jet away at high speed, while at other times, the funnel ejects the water gently, so that the animal seems almost to meander.

The funnel can also be used to blow away the sand or mud on the seabed to find and catch prey that might be lurking below. The octopus female uses the funnel to keep her eggs clean. The funnel is also the steering system. It's movable and can be aimed in many different directions, helping the animal complete tasks like swimming either backward or forward.

Cephalopods usually have three hearts that pump blood and oxygen through their bodies. In highly active cephalopods, like fast-moving squid, these hearts—a main, central heart and one near each of two gills—must sometimes pump very hard to keep enough oxygen in the animals' tissues.

Cephalopods have copper-based blue blood, instead of red blood. Human blood is red because our hemoglobin contains iron. The iron in our blood binds with oxygen in our lungs, then carries it to our muscles. Cephalopods do not have hemoglobin and do not rely on oxygen in this way. Instead, cephalopod blood uses copper to carry oxygen. In some ocean environments, copper can carry oxygen more efficiently, but in other environments, and particularly out of water, copper is not as good as iron at getting lots of oxygen to active muscles. This helps explain why cephalopods sometimes lack endurance: The copper in their blood doesn't get enough oxygen to their muscles quickly enough.

To me, the most fascinating thing about cephalopods is the brain. Some of the cephalopod brain is wrapped around the throat. Like the human brain, the cephalopod central brain sometimes has various lobes dedicated to specific functions, like processing experience and making memories. We also share with the cephalopod the same basic brain cell—the neuron.

But there are also striking differences. Our brain is highly centralized, located of course in our head and protected by our skull. We have a spinal cord, which runs from the brain to about halfway down the backbone. Other nerves run from the brain and spinal cord to the rest of our body, allowing us to control our arms and legs. We have some nerves that respond to an experience without checking in with the brain first. That's how the knee jerk occurs. But in general our nervous system relies literally on top-down control, on commands that come from the central brain.

The cephalopod brain is quite different. It's much more decentralized and seems to have a lot more opportunity for knee-jerk-like responses. Roughly three-fifths of the cephalopod brain resides not in the central system but in the arms and tentacles. This makes cephalopod arms weirdly independent. Arms and tentacles, at times, seem to be able to make their own "decisions." If an arm separates from the body, which might happen for any number of reasons, it can continue to function for many hours. Does the arm "know" what it's doing when it acts independently after being separated from the body, or is it behaving according to some preprogrammed autopilot arrangement? We're unlikely to know the answer to that anytime soon.

Squid and cuttlefish feeding tentacles are usually much longer than the arms. These feeding tentacles can strike with the speed and force of a projectile. There are cuttlefish feeding tentacles that are capable of shooting out and capturing prey in about a hundredth of a second. The giant squid may not enjoy such speedy tentacles, but it may not need the speed. Giant squid have some other pretty impressive tools, instead. Their tentacles widen into paddlelike clubs at the ends, on which are rows of enlarged suckers on flexible stalks. Each sucker is ringed with hard, sharp teeth that embed in the flesh of the prey to grasp and shred the victim's skin and flesh. At the end of the feeding tentacles of the colossal squid are about twenty-five large swivel hooks, each set into a sucker, used to snare prey.

Cephalopods live in all the planet's oceans, except for the Black Sea and the Baltic Sea. As a group, they occupy all ocean depths, although individual species may be more restricted in their movements through the water column. When cuttlefish gave up their protective shells, they evolved a cuttlebone, an elongated and firm structure that is somewhat like a backbone but remains rather rigid. You may have seen a cuttlebone in a bird cage, where it provides calcium to the bird.

A cuttlefish

Cuttlebones, with their honeycomb-like interior structure, provide buoyancy. Cuttlefish can increase or decrease the amount of gas in the cuttlebone, thus allowing the animal to rise and fall in the sea. But cuttlebones are brittle and are destroyed when water pressure is too high. Cuttlefish therefore do not descend into the deep sea.

Octopus and squid species live at many different depths. The ghostly, translucent, and apparently blind deep-sea octopus, *Vulcanoctopus hydrothermalis*—the "hot-water volcano octopus" or the "deep-sea vent octopus"—lives many thousands of feet below the surface around vents on the ocean floor that spew hot water. Only a few inches in length, it eats the strange crabs, shrimp, and other highly adapted fauna that swarm around the heat there. Other octopus species are adapted to shallow waters and a few are known to leave the water and crawl over land when hunting.

Many squid species can navigate safely through a variety of depths, adjusting their physiological responses accordingly. These squid migrate nightly from several thousand feet below all the way up to the sea surface to feed on the variety of marine life that makes the same once-a-day up-and-down trip. The giant squid may also migrate up and down, but no one knows for sure, since the animal's lifestyle remains mysterious.

In fact, scientists know little about the behavior and lifestyle of most cephalopods. For example, we know that many species rest, but we don't know whether cephalopods actually sleep, like mammals and birds, and we certainly have no idea what kind of dreams such animals might have. In between hunting forays, octopuses spend a great deal of time in their dens, their temporary homes. The bobtail squid and many other shallow-water species spend much of the daytime burrowed into the sand. But are they actually sleeping as we would, in the sense that they're rejuvenating their brains? Are their brains processing events into memories, as we do? Or are they in some other kind of neutral, inactive state?

· : · : · : · :

Scientists are trying to answer at least a few of these questions in their study of one of the ocean's larger squid, the Humboldt squid, *Dosidicus gigas*. At up to six feet in length from mantle tip to arm tip and weighing up to 100 pounds (but usually much less) and with a very powerful mantle muscle, this squid is a member of the flying squid family. It has large, muscular fins for swimming and ranges through the ocean, both horizontally and vertically, in schools of sometimes as many as a thousand animals. It eats smaller fish, mollusks, and, sometimes, other squid, including its own schoolmates.

The species has a formidable reputation. Mexican fishermen call it *Diablos Rijos,* or Red Devil, alluding to the numerous stories told by fishermen and others who claim that if a person falls overboard into a school of these animals, only the skeleton will be left by the time the body reaches the seafloor.

Maybe so; maybe not. Scientists are divided on how dangerous *Dosidicus* might be. The Smithsonian's Clyde Roper was bitten on his inner thigh, near his femoral artery. The bite penetrated his diving suit. On the other hand, *Dosidicus* expert Bill Gilly of Stanford University says he's swum with these squid without protection and not been bothered. The solution to the disagreement could be that the Humboldt, like most sophisticated animals, has a flexible temperament. In fact, I think of the Humboldt as being a kind of saltwater version of the coyote, an opportunistic predator who can survive in deserts, on the Great Plains, and on the golf courses of Cape Cod.

Scientists have recently begun to study the Humboldt in depth, because, like the coyote, the Humboldt has begun expanding its range. The species once seemed limited primarily to South American and southern North American salt water, but over the past decade, the Humboldt has become common in coastal waters as far north as Alaska.

No one knows why.

CHAPTER ONE

A SALTWATER SERENGETI

An ocean without its unnamed monsters
is like a completely dreamless sleep.

—JOHN STEINBECK

J ulie Stewart cradled her research subject in her arms. Her ponytail dripped salt water. The back straps of her luminescent yellow waterproof Grundens were twisted tightly to better fit the slight frame of her 5'3" body. She was covered in squid ink. The sun had long since set. It was mid-November, 2009, about a month away from the Winter Solstice. The sea was a bit rough. The air, a bit chilly.

Julie was kneeling on the dive platform of an unnamed government research boat on Monterey Bay. She was outside the safety of the deck rails. Above her were all the stars in the universe. Two miles below her was the bed of one of the world's most sublime kingdoms, the Monterey Submarine Canyon.

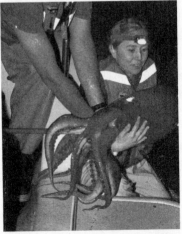

Julie Stewart with a Humboldt on the dive platform

You could have said she was ethereally poised between heaven and earth, but you'd have been taken down a notch if you'd looked around at water level: Monterey Bay is rimmed by all the inglorious mundanity of twenty-first-century human existence. Nearby were the mansion lights of Pebble Beach, the golfers' Mecca. A corporate jet flew overhead preparing to land at the local airport. Off in the distance the towering stack lights of the region's hulking electric plant glistened and beckoned nearly as powerfully as the stars overhead.

Nevertheless, Monterey Bay is a wild place, filled with whales and sharks and shoals of fish and forests of kelp and 20-pound sea slugs and 50-pound Humboldt squid and diaphanous 100-foot-long siphonophores, jellyfish-like creatures that form nets with their poisonous tentacles and wait for prey to come their way. This cold, deep world throbs with energy. It's a saltwater Serengeti.

Rocked by three- and four-foot waves, Julie held her animal close to her chest. At twenty-eight, she was chief scientist of this evening-long research cruise and was part of an informal international team of scientists stretched along the west coast of South and North America. They were all focused on learning more about the biology and behavior of the suddenly prolific Humboldt. The scientists had endless questions. Where had the species come from? Why were these squid here in Monterey in such great numbers? Where were they going? Had something changed in the Pacific that had suddenly opened up a niche that these opportunistic animals were exploiting? Was it because the oceans were warmer? Was it because many of the sea's top predators like whales and sharks had disappeared? Was their explosion in numbers a symptom of some kind of extinction event that affected other kinds of animals but not cephalopods? Or was the species' sudden increase simply one example of the normal long-term ebb and flow of life in the ocean?

On the boat, Julie was at the center of organized chaos and exultant bedlam. Encircling her were five men furiously pulling up five- and six-foot-long squid. Earlier in the season, the team's Humboldt hunts had come up with zilch, but on this particular night the men could hardly keep up with their work.

A huge school of ravenous squid had surfaced just at dusk and were lured to the boat by the large glowing lights on the two-foot-long fishing jigs. In the frothing waters surrounding the boat, squid swirled everywhere. If one squid was hooked on a lure, others saw its vulnerability and attacked.

Scientist John Field pulls in a Humboldt

Cannibalism is common in this species. On another Humboldt fishing trip months earlier, Tom Mattusch, captain of the 53-foot fishing charter *Huli Cat*, pulled up a Humboldt from about a thousand feet down. When he got the squid above the surface, he saw a lot more than just eight arms and two tentacles. He thought at first he'd somehow caught two animals on one lure, but then he took a second look. Most of the body of the first squid lay in the clutches of the second, which was shredding the first animal and eating it.

On Julie's Humboldt expedition, the men were using stand-up rods, about the size used for bluefin. On the end of their 50- and 60-pound test lines were the specially made heavy jigs that the animals proved unable to resist. As squid after squid was pulled on board, the deck was chaotic.

A few feet away from Julie was her doctoral adviser and laboratory head, neuroscientist-turned-naturalist Bill Gilly of Stanford University and Monterey's Hopkins Marine Station. Over the past several years, Gilly had become somewhat of a television personality, having been featured in numerous documentaries about the "dangerous" Humboldts. Some of these documentaries featured the Humboldt as a "killer," the way wolves were once featured as deadly in children's stories like "Little Red Riding Hood." Gilly and his team sometimes roll their eyes at this kind of dramatization.

Gilly was frustrated at having lost a squid after expending quite a bit of energy hauling it up from a thousand feet down. He chewed pensively on a bit of raw tentacle. The squid had escaped the scientist, but this tiny bit of living flesh had broken off the animal and stayed behind, caught on the strong, sharp spikes of the jig. As Gilly gnawed on the squid's body tissue with its still-flexing suckers, he considered the taste. "Not too bitter," he said.

He also considered the temperament of the Humboldt, which he believed to be much more benign than television shows liked to let on.

Gilly has swum with the animals several times without protective gear, in only snorkel and T-shirt. "One of them just came up right at me, took an arm and touched my hand, and went away. If you're kind to them, they'll be kind to you," he told me later. Maybe so. I could see his point. The animal in Julie's arms didn't look dangerous. On the other hand, I wasn't planning on swimming among them.

Not far from Gilly was Rob Yeomans, bending over the open transom in the stern of the boat, pumping the line of his boat rod. Dressed in a hooded black sweatshirt and orange foul-weather pants, he braced his 5'4" frame against the roiling sea. He was cackling with excitement. Rob looked like a metronome, moving forward and backward, forward and backward, pulling squid after squid out of the water. There was a great deal of joy on the boat that night.

At thirty-seven, Rob is a fervent and irrepressible high school marine biology teacher from Newburyport, Massachusetts. By heritage and by emotional makeup, he is a commercial fisher-man, but because of overfishing in the North Atlantic, he was forced to change jobs. I had first met Rob a half-year earlier when I'd attended a Humboldt squid dissection in his high

CHAPTER TWO

school classroom. The frozen carcass had been shipped from Gilly's West Coast lab across the continent, then thawed on Rob's worktable. Rob had wanted to meet Gilly in person, visit his lab, and go out with him to catch some squid. I decided to travel along, to see what all the hoopla was about.

· : · : · : · :

The Monterey region was once sparsely populated with bandits' cabins and shoreline squatters' shacks where Chinese fishermen dried squid for export to distant cities. But the peninsula was "upgraded" by the Pacific Improvement Company at the end of the nineteenth century, when the railroad and improved highways created the new industry of tourism. The real estate developers turned the place into a posh resort with first-class accommodations that included a huge hotel with acres of gardens called the Del Monte, Pebble Beach's golf course, and a clubby lifestyle. All kinds of celebrities showed up, from the Surrealist Salvador Dalí to the Hollywood personality Bob Hope. Teddy Roosevelt galloped his horse along the bay's dramatic shoreline, and President William McKinley visited only months before his 1901 assassination.

Despite the upgrades, Monterey's waterfront continued to smell of rotting fish. The town government established an official "permanent smelling committee" to fine or arrest people who perpetrated offending aromas, but even that didn't work. "Cannery Row in Monterey in California is a poem, a stink, a grating noise," wrote John Steinbeck in his 1945 novel *Cannery Row*. "The canneries rumble and rattle and squeak until the last fish is cleaned and cut and cooked and canned. . . ."

But eventually, the canneries closed down because the region was fished out. Today called Ocean View Avenue instead of Cannery Row, the street looks much like the main tourist street in Provincetown on Cape Cod, or like Main Street in Bar Harbor, Maine, or like any other reclaimed "quaint" fishing town. Steinbeck's bums and whores and slightly seedy, rather rowdy intellectuals and street cops who enjoy a good drink now and then are mostly gone, along with the infamous stink.

In 1992, Monterey Bay became a federal marine sanctuary, which is something like a national park, although not as commercially restrictive. This good fortune has greatly benefited the sea life in the bay. The kelp beds are healthier. Sea otters put on a show for anyone who cares to walk by the seashore and peer down into the waves. And if you don't feel like exploring even that much, you can just see them in the aquarium. The key to continuing the improvement is science—which in this case means the methodical discovery of how to fit the puzzle pieces together. This is particularly challenging because no one knows what the final picture—a healthy ocean ecosystem—should look like.

We do know a few interesting things, though. Scientists have recently found, for example, a very cool ecological chain reaction: The kelp beds that cradle sea life depend on the sea otters who wrap themselves in the tops of the kelp fronds when they sleep. The otters eat the sea urchins which in turn eat the kelp. Too few otters means too many sea urchins. Too many urchins means too little kelp. So it turns out that the otters, who might at first glance seem to be somewhat harmful to kelp by wrapping themselves in the fronds, are in fact enabling the kelp to thrive.

Nature is funny that way. Sometimes the truth is counterintuitive. It took us a while to unravel the otter-urchin-kelp jigsaw puzzle, and when we did, we felt pretty smart. But when we take a step back, we see that the puzzles we are able to solve in ocean ecosystems are only tiny achievements, like puzzles for toddlers with only three or four pieces. "They're not even two-dimensional jigsaw puzzles we have to solve," Gilly said. "They're three-dimensional. We need some kind of systems approach, but we don't even know what that would begin to look like." The ocean, after all, is not about stability but about flux. Change is normal. Everything is changing. All the time. It will be decades and decades before we can understand the sea in any kind of meaningful way.

Among the current, pressing puzzles in Monterey Bay is the sudden proliferation of Humboldt squid. This is not the first time that Humboldt squid have shown up in the bay and elsewhere on the West Coast, but their numbers this time may be greater than in the past. "We don't know whether they're going to stay long this time, either," Julie told me. "But we're trying to at least understand *why* they're here whenever they *are* here." It's a tough task.

The fact that Humboldts have suddenly become common during the summer and fall is not necessarily a bad thing. Charter fishing boat captains like these very large squid because they're not as challenging as game fish to catch, and on the best nights, they're more than plentiful, so most clients come away happy. In the summer of 2009 so many Humboldts swam near the wealthy town of La Jolla that several mass strandings occurred, littering the popular swimming beaches with rotting squid carcasses.

A Humboldt stranding on a California beach

In earlier eras, the public might have been displeased to have their beaches defiled by writhing squid tentacles and slowly rotting squid flesh, but times have changed. Gilly lab doctoral candidate Danna Staaf was in La Jolla when the stranding happened. She went over to the beach to take a look and decided to do some cheap research. It costs money to take a boat out to catch squid, but if the squid come to you, the price is just the cost of a few freezer storage bags.

Strandings of sea life are fairly common. On Cape Cod, sea turtles commonly strand after fall storms. Whales often stranded on the Cape during the Pilgrim era. Even jellyfish strand in huge numbers, making beaches treacherously gelatinous until predators, like gulls, carry the carcasses away. Sometimes animals strand because they're ill, or because they've been trapped in currents, but most of the time, scientists have no idea why marine life washes up on beaches in great numbers.

Standing on the La Jolla beach during the summer of 2009 and looking at the putrefying squid bodies, Danna wondered if the Humboldts had stranded because they suffered from domoic acid poisoning. Domoic acid is a toxin produced by a small subset of algae. Fish and some mollusks are immune to the toxin. When they eat the algae, they accumulate large amounts of the toxin but are not affected.

Higher up in the food chain the situation changes. The birds, mammals, and humans that eat the toxic fish and shellfish suffer greatly. The toxin affects their nerves. In humans, in large enough amounts domoic acid causes sometimes permanent short-term memory loss. In 1987 domoic acid that had accumulated in Prince Edward Island mussels killed three people. More than a hundred others became seriously ill. This also happens on the West Coast. In 1961, hundreds of birds, shearwaters and gulls, began behaving erratically and dropping inexplicably out of the sky in Capitola, California, seemingly attacking the town. People were terrified. Alfred Hitchcock, at work on his film *The Birds,* took note.

In recent years, domoic acid poisonings on the West Coast appear to have become more frequent. To see if the Humboldt strandings might be another example of this trend, Danna took out her knife and began carving out the stomachs of the dead squid. She put each in a separate plastic bag in order to send them to a lab for analysis. Perhaps the stomach contents would provide a clue as to why the animals had died. She was soon surrounded by curious adults and kids.

Staaf explained that the basic body plan of squid and octopuses is quite different from our own. Our own legs and arms evolved from the fins of fish, our distant vertebrate cousins. But the Humboldt squid have very different fins, which have no bones but which are able to propel the animal through the water at top speed.

Curious people, adults as well as kids, peppered her with questions. She ended up giving out sucker rings "like candy" to those who wanted them, which was, surprisingly, almost everyone. She gave one kid the squid's beak, but he returned later with a glum expression. It turned out that his mother had decreed that this prize of war had no place in the family automobile.

Then she showed some children some packets of sperm, called spermatophores.

"See how they pop open in my hand? Isn't that cool?" she said.

"What's sperm?" asked one kid.

Maybe that's not the best direction to have gone in, she thought to herself. But she decided that the question required an answer.

"Well, it's what mixes with eggs to make baby squid," she said. It sounded kind of like a recipe for baking a cake, but it did the trick.

A few days after sending her frozen squid stomachs to the lab for analysis, Danna heard the results: no domoic acid. She wasn't disappointed because she felt like she was part of an international team of scientists gathering as much basic data on the Humboldt as possible. Her negative result had had a positive effect by providing one more small piece of information that would help scientists understand the whole picture.

· : · : · : · : ·

Humboldt squid stayed in the headlines for much of 2009. Even PETA—People for the Ethical Treatment of Animals— got into the act. The organization took advantage of the press attention by posting a banner on one beach that said: "Warning: Predator in the Water! You! Go Vegetarian!" PETA's plea didn't work: Humboldt squid sandwiches are now available at many seaside restaurants.

BLUE BLOODS

The absence of a skeleton in a marine life form constitutes a form of perfection.

—JACQUES-YVES COUSTEAU

On the Monterey Bay research boat that November evening, Julie Stewart continued to cradle her research subject. She was waiting for the precise moment to ease her five-foot Humboldt, fins first, into the rough waves. If she made a mistake or just dropped the animal onto the sea surface, the squid might have trouble swimming away. Or, disastrously, the $3,500 satellite tracking tag she had attached to the fin might come off.

She bent down closer to the water. She might have found herself *in* the water but for John Field, a 6'3" surfer and research biologist with the National Marine Fisheries Service. John grasped Julie's life vest. From the safety of the boat deck, he braced himself and held tightly, stabilizing Julie so she could concentrate on the task at hand. The mantra at sea is "one hand for yourself, and one for the boat," but she needed two hands to hold the animal.

For an instant the Humboldt, with its strange baseball-size eyes, looked directly at Julie, as though trying to cross the gulf of 700 million years of evolution. The animal flashed red and white, red and white, showing off its chromatophores.

"Kind of like a disco," Gilly commented.

We experience the Humboldt's show of red as a display of anger. Maybe our own brains are hardwired to make the connection between the color red and the flow of blood. But that's not why the Humboldt turns blood-red. In the ocean the color red disappears quite quickly because its long wavelengths cannot easily penetrate water. What appears red above the water appears merely dark below the surface. When the Humboldt turns red below the water surface, it is making itself invisible.

No longer buoyed by salt water, Julie's Humboldt was in fact rather helpless. Its eight arms and two feeding tentacles were pulled down by the full force of the earth's gravity. It was not accustomed to the sensation. Out of its medium, its behavioral choices were limited. A fish out of water flaps on the deck of a boat, trying to escape. A squid, however, lacks the framework of a skeleton. It has no bones or any hard internal structures, other than a flimsy "pen," the evolutionary remnant of the shell that all cephalopods once had. (The pen is so named because it reminded people of ink-filled quill pens.) Made of material somewhat like your fingernails, the pen is easily bent and substitutes for a backbone, and as is the case with our back-bone, many of the squid's muscles attach to the pen.

But the flimsy pen can support these muscles only when the squid is in the water. When it's beached or hauled on board a boat, the pen isn't particularly helpful. As a result, the Humboldt cannot flop around like a fish. What it *can* do is slash with its beak, which is quite sharp. A nasty wound is not uncommon. The animal's arms and tentacles are also dangerous. "The tentacles are their secret weapon, their jack-in-the-box surprise," Gilly said. "If the teeth on the arms get you, it's like getting bitten by fifty garter snakes."

· : · : · : ·

One reason why Julie's squid lay so passive in her arms may have been related to the animal's blood, which supplies oxygen to its cells using chemistry that's quite different from our own. "They have blue blood, ice crystal blue," said Gilly, "as blue as an iceberg."

Blood, of course, flows through an animal's circulatory system, carrying oxygen to all the cells of the animal's body. Oxygen, the third most common element in the universe and essential to life, is produced by land plants, but surprisingly, most of the oxygen in our atmosphere is produced by marine algae.

It's a good thing these algae are around. We owe them our very existence. Were it not for them, we would asphyxiate. Living cells need to have a constant source of oxygen. In vertebrates, oxygen enters the body through lungs and clings to the iron in a hemoglobin molecule. The hemoglobin then travels through our circulatory system, bringing oxygen to cells that need it. If we're running, for example, the hemoglobin drops off extra oxygen to our leg muscles. Not all animals, however, use hemoglobin. Some animals, spiders and lobsters for example, substitute a compound called hemerythrin for hemoglobin.

Mollusks and many other marine animals use hemocyanin— a molecule that may have evolved as long as 1.6 billion years ago, long before the first mollusks and roughly a billion years before the Cambrian Explosion. This seems pretty long ago to me, but scientists interested in molecular evolution believe that hemoglobin, our own oxygen-carrying molecule, may be even older, perhaps even dating back four billion years, to just after the time the earth was formed.

Even if the mollusk's hemocyanin was not the first oxygen-carrying protein to evolve, it must have done its job fairly well. In the Ordovician, the period following the Cambrian, cephalopods proliferated. The Ordovician was a rather eccentric period in earth's history: Most of the Northern Hemisphere was under water and most of the planet's land was gathered into one single supercontinent, Gondwana. This southern supercontinent was slowly drifting over tens of millions of years, inching its way relentlessly south.

For a while, the seas were deliciously warm and the planet seems to have been a kind of Garden of Eden, a time of nirvana that allowed life to flourish in many different forms. A few cephalopod species grew large enough to rank among the largest animals then extant. Protected by long, straight, conical shells, with numerous arms poking out and dangling below their eyes, the larger cephalopods were quite fierce. One group, *Cameroceras*, lived in shells that may have been as long as 30 feet—the size of a large RV.

Cameroceras, which we would easily recognize today as a cephalopod despite its burdensome shell, was certainly formidable. It may well have been the ocean's top predator. But it would not have been very maneuverable. For most of the Ordovician, this probably wasn't a drawback: Since life proliferated in the warm shallow seas, all *Cameroceras* had to do was hang out just above the seafloor until some tasty morsel passed by.

During this period, cephalopods ruled. Unfortunately for them, nothing lasts forever. Circumstances were about to change. Gondwana continued its southward journey. As the supercontinent headed nearer the South Pole (eventually North Africa would be directly over the pole), the climate chilled. Gondwana glaciated and the world became cold. This may have happened relatively abruptly, over a period of only several millions of years. Ocean life had little time to adapt and many of the planet's species, including many cephalopod species, died off.

The glaciations themselves may have tripped the climate change, but other explanations have also been offered. One NASA researcher has suggested that a very powerful explosion of a star, a gamma ray burst, may have occurred near enough to earth to destroy the protective ozone layer. Whatever the cause, the cephalopods as a group once again managed to survive. No one knows for sure why cephalopods are so resilient, but their ability to survive might be due in part to their use of hemocyanin in lieu of hemoglobin.

· : · : · : · :

Fast-forward to the Mesozoic era, the era of the dinosaurs and the Triassic, Jurassic, and Cretaceous periods. From about 245 million years ago to 65 million years ago, cephalopods once again ruled the seas. But this time they did not rely on size and power, since they certainly couldn't compete on the same scale as large oceanic predators like the 50-foot Mosasaurs, marine lizards that slithered snakelike through the oceans hunting, among other prey, cephalopods; or like the 500-pound, 10-foot-long sea turtle *Protostega* that patrolled shallow waters relentlessly in search of luscious squid lunches.

In the face of such enemies, the cephalopods for the most part opted not for size, but for sheer numbers. The predominant cephalopod group, ammonites, spread everywhere throughout the planet's oceans, although they seem to have preferred shallow coastal seas. We know this today because their fossilized shells have turned up in the oddest places—in mile-high mountains in Afghanistan, all over the American Midwest and Southwest, and layered in the southern cliffs of Britain, along the English Channel.

An ammonite fossil

Ammonite fossils were so common around the English town of Whitby that the town's early coat of arms showed three of them. Only these three had been slightly adulterated to meet the needs of the local belief system. Early on, the people of Whitby had decided that the ammonites were the remains of coiled snakes, and a local legend had evolved about a saint named Hilda who rid the town of snakes by turning them to stone.

Of course, the people of Whitby never actually found any ammonites with snakes' heads. So, to validate the legend, they fabricated the evidence: They carved snake heads onto the ammonites, then claimed said heads had always been there.

For a time, Whitby remained fairly committed to the tale of St. Hilda. The town shield featured ammonites with snake heads. But finally science ended the fun by explaining that the coiled fossils were not the remains of snakes, but of animals that had long since disappeared from the earth. The ammonites are still on the town shield, but the snake heads have disappeared.

Of course, ammonite fossils are only the shells in which the animals lived. Science knows almost nothing about the cephalopod that occupied the shells. Curiously, we have more fossilized soft parts from earlier nautilus species than we have for the ammonites, despite their proliferation, so we're not quite sure what the animal inside the shell actually looked like, but scientists extrapolate from modern cephalopods to suggest that ammonites, also, had well-developed eyes, a raspy radula, and many tentacles.

From about 240 million years ago until 65 million years ago, ammonite species were so prolific, and sometimes evolved and disappeared with such rapidity—in the blink of an eye as geologic time goes—that they have become important signposts worldwide for geologists trying to age a particular rock stratum. They are a central pillar of the science of biostratigraphy—the science of correlating ages of rock with the fossils of extinct animals found in those rocks. Some ammonites evolved, proliferated, then disappeared in only one or two million years. If geologists find ammonite fossils in a rock layer, they can age a layer of rock quite accurately to within a million years or so. This can be done worldwide, so a layer of rock in China may be connected to a layer of rock in the American Southwest or in Britain just because the same fossilized ammonite species appears in all three places.

In the eighteenth and nineteenth centuries, ammonites also helped people wrap their minds around the difficult demands of imagining both geological timescales and evolution itself. In Europe in those days, collecting ammonite fossils was a quite respectable outdoor occupation. Even women were allowed to participate. Most ammonites are small and can easily fit into a pocket or purse, although a very few shells may be five or six feet in diameter. Amateur collectors couldn't help but notice that the various ammonite species appeared and then became extinct in correlation with specific geologic layers of rock. Charles Darwin certainly wasn't the first person to notice this, although he was the first to place this interesting little factoid into an overarching theory.

When the dinosaurs died out 65 million years ago, the ammonites also became extinct. But again, the cephalopods as a group survived. We know about the proliferation of ammonites because of their fossilized shells, but we know very little about the early shell-free species. Fossil evidence of their soft bodies is rare, but from time to time, fossilized cephalopod ancestors do turn up. In the summer of 2009, paleontologists discovered a 150-million-year-old squid, an animal that would have shared the seas with the ammonites. Found in Britain, in a region well known for the quality of its fossils, the squid was easily recognizable. Its inch-long ink sac was so well preserved that scientists were able to take a sample of the ink, grind it up, add some liquid, and then use that very ink to sketch the fossilized animal.

Sketch of a 150-million-year-old squid fossil

At about the same time, other scientists reported finding a 95-million-year-old fossil of an octopus in limestone deposits near the present Mediterranean shoreline in Lebanon. This animal, too, lived in the ocean while the dinosaurs still thrived. It also had a distinctly preserved ink sac. Scientists were amazed by the fossil's overall quality, which showed an octopus that looked quite like today's modern octopus. Since not much has changed in the octopus's basic body shape, a few marine biologists believe that the octopus may be an evolutionary dead end and that there aren't going to be many more mega-design changes.

∙ ∶ ∙ ∶ ∙ ∶∙

With the satellite tag activated, Julie waited for the waves to settle. At last, after several minutes, the boat rocked into position. She slipped her squid back into the bay, gently, like a mother laying an infant in a cradle. She felt its rough, craggly skin against her fingers. The loose texture made her wonder whether the animal was older than the other two she had already tagged that evening. It certainly was much bigger, almost as long from mantle tip to feeding tentacle tip as Julie was tall. It was many pounds heavier than most of the roughly 50-pound Humboldts that routinely turn up in Monterey Bay.

It was 7 p.m. The cruise had started just after 4, and already Julie had the last of the three research subjects she'd hoped to tag. She and Gilly hoped the expensive computer chips inside the tags attached to the squids' fins would yield some useful information.

The tracking tag

The team wanted to know where the squid traveled. The daily lives of animals—even animals living on land—remain mysterious. We know a tiny bit about charismatic megafauna like whales and elephants and lions, and we're fascinated by the sparks of intelligence shown by dolphins and chimpanzees, but we're pretty ignorant of the habits and preferences of much of the animal life that surrounds us. Learning about animals has been one of humanity's greatest adventures. Each little step we take that advances our knowledge—"Whales sing to communicate with each other" or "Chimpanzees work together and use tools like sticks to acquire food"—feels like the discovery of a new universe to us. Shortly before his death in 1987, sea turtle biologist Archie Carr stood on a Florida beach and spread his arms wide, as if trying to embrace the whole of the Atlantic Ocean. "Where do they go?" he asked about the turtles that had become his life's passion. He wasn't asking for himself. He was leaving a research question for the generations of marine biologists that would follow. Today, in large part because of Carr's passion, we know a great deal about where sea turtles go in the sea, about what they eat, and about how they navigate their way back to the beaches where they hatched.

But our understanding of the behavior of these few sea species is anomalous. Of most sea life, we know nothing. Indeed, much of the life in the ocean has yet to be catalogued. Discovering facts about animals that live in the ocean depths is inordinately difficult—expensive and time-consuming and technology-dependent. Money is tight. We can't afford to spend much on each individual species down there. But, to Julie's good fortune, some money at least is available for studying Humboldts. Commercial fishermen charge the Humboldt squid with the crime of eating salmon and hake and smaller squid, species that commercial fishermen sell at market. This connects the Humboldt to a big-money product and so makes research funding more available than it would be otherwise.

· : · : · : ·

As her third tagged Humboldt swam away, Julie was thrilled. So was Gilly. "We've had hauls like this down in Baja," he said, "but never anything like this up here before."

For a scientist, data is the be-all and end-all, the ultimate goal, the sine qua non of fieldwork. No data, no science. No science, no funding. The goal of an evening cruise like this is to get enough information to keep Gilly's lab humming for months. It doesn't always happen. Fishing for data is as risky as fishing for big-money bluefin. You might hunt and hunt and just as easily come up with nothing as come up with a fortune. The odds are better than wasting your time in Vegas, but not by much.

Julie's tracking tags were fairly large in size, 175 millimeters (a little less than 7 inches in length) and 75 grams (a little less than 3 ounces)—"the size of a karaoke microphone," Gilly mused. You might use something about the same size on a sea turtle or a tuna or a shark. The tags, called Pop-up Archival Transmitting Tags, come with a pair of plastic pins, but it's up to the scientist to figure out how best to attach the tag to the research animal. The scientist can also program the tag to release from the animal in a specified number of days. Stewart had chosen to attach the tag to the fins, using the pins, and to .

program the tag for release in seventeen days, by which time the data storage chip would be full.

As the squid moves vertically and horizontally through the water, the computer chip in the tag records information, including temperature and light levels, from which depth can be calculated. This information is recorded on the computer chip, but not all of that is sent to the satellite. Instead, because transmitting the data to a satellite is expensive, Julie has opted for the information to be sent to the satellite only periodically. From the satellite, the data is sent to her laptop.

The receiving satellite, one of a six-satellite system called the Argos System, has been providing scientists with important animal behavior data for more than thirty years. Today, well over four thousand tagged animals worldwide provide data via this technology. Much of what we know about sea turtles, for example, comes from Argos technology. By using tracking tags, Barbara Block, a colleague of Gilly at the Hopkins Marine Station who studies sharks, learned that great white sharks migrate far offshore into the Pacific, overturning the belief that the animals stay fairly close to the shoreline. Other tracking tags have shown that dolphins dive much more deeply after prey than hitherto expected. Recently scientists began tracking walrus migrations through the Arctic seas.

The information from the tag that's beamed up to the satellite then down to the scientist is useful, but the information archived in the tag itself, the instant-by-instant story of what the animal's been up to, is the real treasure. When the tag pops up, it transmits its location to the satellite. Scientists will go to great lengths to retrieve that tag, since it has more of what they want. But they also want the instrument itself, since it can be sent back to the manufacturer for reprogramming and reuse. Most marine labs can't afford to waste $3,500.

Unfortunately, looking for a tracking tag about the size of a karaoke microphone bobbing in the waves of the ocean is like looking not just for a needle in a haystack but for a needle in a *moving* haystack. The task can be both time-consuming and frustrating. You know the item is there, but you just can't see it.

Stewart remembers being out on the ocean looking for a tag and knowing from the satellite signal that it was right there, almost beside her. But she just couldn't find it. Eventually she had to give up and accept the financial and scientific loss.

Most tags carry information about a reward if found. Scientists often get them back that way. Fishermen know to pull things like that out of the water. Beachcombers may pick them up. Surfers may find them. Salvador Jorgensen, a great white shark researcher in Barbara Block's lab, once searched high and low for one of his tags. Determined to get his data, he followed the pinpointing signal. It led to a residential neighborhood, then to an individual house. He knocked on the door.

"Do you have my tag?" he asked.

It turned out to be in the wet suit of a surfer who had found it in the water, put it in his pocket, then carried it home and forgot about it.

· : · : · : · :

If following the animal can be expensive, every once in a while, scientists get lucky. The animal comes to them.

Architeuthis ON ICE

And God said: Let the waters teem with countless living creatures.

—GENESIS

T he morning of June 25, 2008, wasn't a good one on
Monterey Bay. It was a little more than a year before
Julie and her research team came upon the huge shoal
of Humboldt squid.

That early summer morning, the shark hunting hadn't been
great. The weather wasn't holding up. The bay was choppy.
The wind was picking up. Conditions were going from bad
to worse, and so, even though it was only around 9 a.m., the
Pelagic Shark Research Foundation's Sean Van Sommeran
decided to call it a day. No shark tagging on this trip. He was
about to head back to Santa Cruz. Then, like Theophilus
Piccot more than a century earlier, Sean spotted something
odd. A large object was on the surface of the sea, rising and
falling in the heightening swells. A number of birds were
battling over something valuable. What Van Sommeran saw
several hundreds yards off seemed to him to have all the
features of a feeding event. Perhaps, he thought, it involved
a shark. Maybe he'd get at least one animal tagged before
heading home. Otherwise, all that expensive boat fuel would
be wasted.

As he drew closer, Sean could see a slick on the water's
surface. Gulls and shearwaters and even a black-footed albatross
were having a field day. He realized the birds were feasting on
some kind of animal, alive or dead. He thought it might be a
marine mammal, perhaps an elephant seal, since the boat wasn't
that far from Año Nuevo Island, a favorite haul-out for the
massive creatures, which can-weigh as much as 5,000 pounds.

As they closed in, the birds drew back. Sean saw a large,
moldering mass just beneath the small whitecaps. Bits and

pieces of torn and shredded white and red protoplasm were still attached to the main body.

Seabirds love eyes. Those delectably soft, gooey organs are usually the first to be eaten. And in this carcass they were indeed gone. The stomach had been torn away as well. Dangling from the main body of disintegrating flesh were ten appendages, none of which seemed intact. Just a bit of what once might have been a fin rose from time to time above the waves.

At first Sean thought the decaying specimen was a *Moroteuthis robusta*, a robust clubhook squid. Said by some to be the ocean's third-largest squid, after the giant and colossal squids, *Moroteuthis* can reach human-size lengths if measured from tentacle tip to mantle tip. While they aren't as common in the bay as market squid, *Loligo opalescens,* neither are they rare. He'd seen many others.

Then Sean took a second look. The specimen, shredded as it was, seemed too big to be a clubhook.

"That might be an *Architeuthis*," he said, thinking aloud. "It's gotta be. It's so big."

"What's that?" some students along on the trip asked him.

Van Sommeran was thrilled. He'd read about the giant squid, but, to his knowledge, there hadn't been any recovered in Monterey Bay. To him, the find was better than a gold strike.

He explained the specimen's value to the students on board: Only a few giant squid had been identified along the California coastline. He knew the teuthologists would want to take a look at it. He regretted the loss of the specimen's eyes. The skin still seemed to be changing colors. He figured that he might have missed finding a live animal—or at least one with intact eyes— by only a few minutes. Still, he knew he was in possession of a scientific treasure trove. His first thought was that the animal might have been killed by a shark, but no one would ever be able to pinpoint the exact cause of death.

Sean picked up a gaff and leaned out over the water, gently pulling the carcass toward him, trying to secure it without further damaging it. Two crew members picked up nets and, together, with part of the carcass in each net, hauled it on board. They beelined back to Santa Cruz.

He started calling his contacts.

"I've got an *Architeuthis*," he said triumphantly.

"How do you know?" one person asked.

Duh, Sean thought to himself. "You don't need a guidebook to know an *Architeuthis* when you see one."

The scientists who met him on the dock confirmed his find. They were as excited as he was. Only four other such specimens had been recorded on the California coast, and none of those had been found recently. Marine science has come a long way since the days of Theophilus Piccot, when the most pressing question was merely to prove the existence of such a huge skeleton-free creature. Modern science had developed a number of analytical techniques that could be applied to the squid, dead though it was, including, of course, DNA analysis. This dead animal was bound to be the center of a lot of scientific attention.

· : · : · : · :

Theophilus Piccot's squid may have marked the beginning of serious scientific research into the giant squid, but our specific knowledge hasn't increased much over the past hundred years. With the invention of deep-sea submersibles, professional and public interest has grown, but some important and rather basic questions remain to be answered. One of the most pressing: "Just what *is* a giant squid?" Incredibly, more than a century after the photo of Piccot's squid appeared in the British science journal, scientists have yet to determine how many species of giant squid, *Architeuthis*, exist in the world.

The word *species* can be either singular or plural. It usually denotes a particular group of animals that can breed and produce offspring, and these offspring can also produce offspring. Whereas a horse and a donkey can produce a mule, the mule is sterile and cannot produce its own offspring. Therefore, a horse and a donkey are considered two separate species.

A species is given two names, both taken from the Latin language and by convention italicized. The first name denotes the animal's genus, a grouping of very close relatives. The second denotes the specific species. The easiest way to think about this, the Smithsonian's Clyde Roper suggests, is to think about a car—a Ford Fusion, perhaps. "Ford" would be the genus; "Fusion" would be the specific species. By convention, the first word of the scientific name is capitalized; the second is not.

Over the past several centuries, whole scientific careers have been based on debating which animals belong to which genera ("genera" is the plural of "genus") and whether two very similar animals belong to the same species or should be classified as two separate species. To the outside world, these debates may seem like medieval how-many-angels-on-a-pinhead debates, but the scientific naming of an animal is more than just esoteric. It is the foundation of the biological sciences.

When scientists talk about various organisms, they have to be sure they're all talking about the same thing. Species may look alike and even seem to the casual observer to *be* exactly the same, but looks can be deceiving. Two look-alikes may turn out to behave quite differently. Each individual species has something special to offer.

Evolution has blessed Julie's *Dosidicus gigas,* for example, with unusual sucker rings, sharp enough to easily slash a person's arm or leg. The squid's beak and sucker rings have the strength and sharpness of well-manufactured steel, yet they're wholly organic. This remarkable fact is providing materials designers with clues as to how to design new substances that remain hard and sharp, like strong metals, but are made entirely of protein rather than of elements mined from the earth. Researchers hope that they will eventually be able to design a material that mimics this wholly organic structure. If they can, the benefits to human medicine will be profound. Some investigators, for example, hope to be able to use materials like this to create organic aids to amputees.

That's why species conservation is important—not only because of conservation itself but also because those species are gold mines of possibility. And in order to ensure the survival of these species, scientists must know as much as possible about their needs. "Each species has slightly different requirements for life and for survival," Roper said. Several species of squid that look the same may spawn at different times, or perhaps one group will spawn and attach its eggs to the seafloor while another will release its fertilized eggs into the water. Or one species may possess the key to curing a human illness while its closely related cousin does not. "You really need to know the biology and the characteristics of each species," Roper said. "In medicine, so many animals, of course, are just extremely important in providing pharmaceuticals. In closely related species, one might be incredibly effective in treating a particular disease, but another that's very similar might be totally ineffective."

A species name also helps bring clarity to some very messy human discussions of the natural world. The common name for an animal varies from language to language and from culture to culture, but the Latin name, the scientific name, is universal. This eliminates confusion.

This is particularly true in the case of the giant squid, with its centuries-old history of so many different popular names for what might well have been the same animal. Since the mid-1800s, the giant squid has been given many different scientific names as well—at least twenty, denoting as many as twenty different species. But few scientists genuinely believe that the ocean has as many as twenty different giant squid species. The confusion primarily stems from the lack of specimens to study and the lack of time and money to do research.

The question is unlikely to be settled officially anytime soon. Eventually, when funding becomes available, DNA analysis will probably be called into play. Meanwhile, some scientists believe there is only one *Architeuthis* species spread throughout the earth's ocean, while others hold that there may be many. Most experts, according to Roper, believe that there are three separate species—one based in the North Atlantic, one based in the North Pacific, and one based in the Southern Ocean, the contiguous ocean surrounding Antarctica.

· : · : · : · ·

Only hours after Sean Van Sommeran brought his tattered specimen on board, Julie Stewart got a surprise phone call from her colleague John Field. The squid was going to be dissected at Field's National Marine Fisheries Service lab on the north coast of Monterey Bay, in Santa Cruz. There would be lots of scientists at the party, among them her doctoral adviser, Bill Gilly, and Julie was invited. Field would be the lead author of the paper that would report their work to the rest of the scientific community.

The next day the *Architeuthis* carcass was hauled out of Field's subzero freezer and laid on the cold steel table. Some of the researchers gathered around the table like a bunch of television doctors performing an autopsy, although jeans were more common than white coats at this surgical procedure. Other scientists, observing officials, television cameras, print reporters, and other hangers-on milled about. Julie found herself in the middle of a scientific and media frenzy. In her casual clothes, without a chance to gather her thoughts, she was called upon to explain her research in front of very demanding television crews.

Word had spread fast. Everyone wanted to be there. Researchers from San Francisco's California Academy of Sciences had driven several hours to attend the gala. When it came time to dissect, Julie pulled on her latex gloves and stepped up to the table. Once again, her hair was pulled back into a ponytail. She dug into the carcass, trying to ignore the caustic smell of ammonia.

She carved out the gills. She wanted to do a comparison between the gills of this giant squid and those of her own long-term research subject, the Humboldt. Her hope was to better understand the details of how squid gills work. When she first thought about what questions to ask for her doctoral thesis, Julie was interested in how Humboldt squid managed to survive for long periods of time in parts of the ocean that had low levels of oxygen.

While the ocean contains oxygen almost everywhere, it is not evenly distributed. Surface waters contain almost as much oxygen as our atmosphere, but levels decrease as you descend—to a point. Then levels begin to rise again. This middle layer, the oxygen minimum layer, is not static. In any one location, like everything else in the ocean, it fluctuates. Over time, the layer may shift up or down hundreds of feet, depending on factors like water temperature.

One of the marvelous things about the Humboldt squid is its ability to move quickly from layers of water with low levels of oxygen to layers with high levels of oxygen, so it has a larger habitat range than many sea species. Some other sea species are able to survive for a time in low-oxygen environments if necessary, but it's possible that the Humboldt and some other cephalopods may sometimes actually seek these places out as refuges, since they can't easily be followed by predators. The sperm whale can dive into these depths to follow and catch squid prey, but only because this whale is able to hold its breath underwater for anywhere from thirty to ninety minutes. Most other squid predators do not enjoy this particular talent and thus cannot survive for long in low-oxygen layers. The ability to survive in water with low levels of oxygen may be yet another reason why cephalopods have survived so many extinction events.

Julie also wanted to know how *Dosidicus* could move so quickly from layer to layer. The species' ability to travel quickly up and down the water column, through various high- and low-oxygen layers, would be somewhat equivalent to our being able to jog from sea level to a 20,000-foot peak without feeling oxygen-deprived. Perhaps, Julie theorized, the Humboldt has particularly large gills and can take in more oxygen than, say, market squid—squid that tend to live closer to the oxygen-rich surface. Julie wanted to compare the Humboldt's gill size to that of other squid, octopuses, and cuttlefishes. Thus, this *Architeuthis* specimen was like manna from heaven. She could include what she learned about the gills of this rarely studied animal in her thesis.

Around the dissection table, the scientists began to divvy up the *Architeuthis* body parts. Field himself regretted not having the stomach to study. His specialty is looking at the stomach contents of animals that live in the sea in order to find out what they eat.

Other scientists regretted the lack of eyes and gonads, which had been picked over quite thoroughly by greedy seabirds. "I guess those are the most delicious parts," Julie decided.

Samples were taken to see if toxins had accumulated in the specimen. The amount of mercury in the body tissue could help scientists figure out where *Architeuthis* sits in the food chain. The more mercury accumulated by the animal, the higher the animal is in the eat-or-be-eaten hierarchy. Predator fish like large tuna and sharks usually have much more mercury in their body tissue than do prey fish like herring. The toxicology results surprised the scientists. The giant squid flesh contained much lower levels of mercury and other toxins than they expected.

Field explained later that "in general, the contaminant levels from the tissue samples all suggested low levels in the *Architeuthis*, which was interesting, as many other animals sampled from Monterey Bay have been shown to have high to very high contaminant loadings." The scientists speculated that the giant squid spent very little time near the Bay shoreline. But since little is understood about how *Architeuthis* metabolizes food, the jury is still out on what the results mean.

Lou Zeidberg of UCLA, who often works with Gilly and Julie on various projects, wanted to examine the specimen for statoliths—minuscule structures made of calcium carbonate, like seashells. Small enough to fit easily on the tip of your finger, these little structures, part of the vestibular system that helps the animal tell up from down in an ocean environment, behave somewhat like the balancing apparatus in our own ears. Fish have similar objects, called otoliths. Statoliths and otoliths differ from species to species. Experts can look at a statolith or otolith and identify the species it came from. "We can also measure them and extrapolate how big the prey species was," Zeidberg explained. "You find these in the squid stomach, and find that perhaps it's 4 millimeters long, and know that it came from a fish that's 40 centimeters long. John Field looked at all the stomach contents of *Dosidicus* and extrapolated the size of the hake that the squid were eating. This way, you can figure out the food web in a much more powerful manner."

Scientists have also learned to use statoliths, which may be retrieved either from a predator's stomach or from the body of a dead specimen, to estimate the age of a squid. Statoliths grow over the course of an animal's life, putting on a small layer each day. Experts can read the layers and age the specimen rather like we can read the age of a tree by counting the rings. It sounds simple, but there's an art to the process. Cutting a statolith for study is like cutting a diamond: Before it will reveal its secrets, it must be properly prepared. This requires delicate craftsmanship, since the three-dimensional, irregularly shaped object may easily be destroyed. Despite its anomalies, the statolith has two basic sides, which must be carefully filed to reveal the tree ring–like lines inside. In a sense, the tiny object must be peeled down to expose the layers. "A lot of people find it helpful to think of an onion as an analogy," Zeidberg said. The scientific craftsmen must be careful to cut at only certain angles so as not to destroy the layers themselves.

Owing to a scarcity of data, how long *Architeuthis* lives is currently a matter of debate. A few researchers think the animal, which lives in deep, cold waters, might live as long as twelve years. Others suggest the upper age limit might be three years. Acquiring and studying a large number of *Architeuthis* statoliths could help resolve the disagreement.

Zeidberg sat firmly in the three-year camp, but he wanted to know for sure. He dug the statoliths out of the *Architeuthis* head, then took them back to Gilly's Monterey laboratory. Two experts in *Architeuthis* statoliths, a husband-and-wife team, just happened to be in town for a conference. Zeidberg joined with them in trying to age the giant squid specimen, but the right equipment wasn't available. The Gilly lab is well equipped with modern technology, but it lacked the precise technology—a very high-powered microscope with certain very specific types of lighting—to prepare statoliths. The couple took one of the statolith specimens back to the Falklands, where their own lab had the appropriate tools. As of this writing, their findings haven't been published.

· : · : · : · :

During the *Architeuthis* dissection, Gilly was in his element. He hadn't been feeling well when he'd been called in, but all the excitement cheered him up. Wearing two pairs of glasses, a regular pair on his nose and a jeweler's visor with magnifying lenses on his forehead, he spoke to the television cameras. "This is a miraculous thing," he said. "Only four or five of these things have been found in the history of California science as far as we know."

Gilly was surprised to find chromatophores—cells that contain color—on what seemed to be interior muscular tissue. Julie had also found chromatophores on the gills themselves. "Unusual," she mused. Squid, octopuses, and cuttlefishes have a wide palette of colors on their skin that they can flash under a variety of conditions, but these are believed to be modes of communication.

"Why," Gilly wondered, "would there be chromatophores *inside* the animal?" There are several other species of squid that also have chromatophores on the inside of their bodies, but no one knows what function those chromatophores might fulfill. Gilly also noted that the specimen's ganglia—clusters of nerve cells—seemed to be proportionally smaller than equivalent clusters in the Humboldt.

"Does that mean that *Architeuthis* is less intellectually advanced than the Humboldt?" I asked.

He said he just couldn't answer such a question. Too little is known about cephalopod intelligence to make such comparisons.

Gilly cut off a piece of flesh and tasted it. For science's sake. Previously Clyde Roper had grilled up some long-frozen giant squid flesh for a dinner with friends and colleagues in honor of one of his students who had just passed his doctoral exams. This was in St. John's, Newfoundland, not far from where Piccot and Squires chopped the tentacles off their giant squid more than a century earlier. When Roper handed his hot-off-the-grill delicacy around to the dinner guests, it turned out that he was the only diner willing to partake. He didn't eat much. *Architeuthis* flesh tasted like ammonia, something like floor cleaner, perhaps, he declared. Since his experiment, other scientists had concurred with Roper's "floor cleaner" finding.

However, Gilly disagreed with that reigning scientific wisdom. He had changed his vehicle's battery terminals more than once in his life and accidentally gotten battery acid on his lips. It wasn't a pleasant experience. The giant squid flesh, he said, reminded him of that.

"But the texture was nice," he said later. "The difference may be in the fact that we had it sashimi-style and Clyde had cooked it. We need to have a group tasting. After all, it's all in the presentation."

· : · : · : · :

One of the most important tasks of the field scientist is to properly preserve specimens for study by later generations. In the past hundred years or so, all over the world, vast libraries of such information have been archived in dusty museum drawers, university basements, and, in modern times, ultra-deep-freeze freezers.

These archived specimens will provide answers to questions scientists don't yet even know they want to ask. In 1835, for example, Charles Darwin collected and sent back for archiving a mockingbird from the island of Floreana in the Galápagos. Several decades later, the species became extinct. Today, scientists are using the genetics from the archived bird to reestablish the species on the island.

And so, what remained of Sean Van Sommeran's *Architeuthis* was put in plastic bags and preserved in a freezer at -20° Celsius. Eventually, it was sent to the Santa Barbara Museum of Natural History, an institution with scientific roots reaching back to the days of Theophilus Piccot. There it was archived on ice by Eric Hochberg, a world expert on cephalopod taxonomy.

· : · : · : · :

Only a few months after Sean Van Sommeran brought his tattered giant squid to the scientific team, a craigslist posting appeared: "Free Giant Squid. Location—the Ocean. Can no longer afford food costs, due to recession. To good home only." There was only one response following the entry: "I would take it in, but I'm not sure if it will get along with our cats."

FUZZY MATH AND TENTACLES

It appears that the tentacles coil into an irregular ball
in much the same way that pythons rapidly envelop their prey
within coils of their body immediately after striking.

—TSUNEMI KUBODERA AND KYOICHI MORI,
2005 *Proceedings of the Royal Society*

On July 17, 1838, American diplomat Richard Rush set sail from London on the triple-masted schooner *The Mediator*. He was headed for the United States with eleven boxes of English gold sovereigns—the dowry for the soon-to-be-consummated marriage between scientific knowledge and the American people. Carrying boxes of money in a ship that could easily sink in a transatlantic crossing was risky, but in those days there was no way to transfer money except to move it physically from one place to another.

The money was a bequest from an obscure English scientist, James Lewis Macie Smithson, the illegitimate son of the Duke of Northumberland. Smithson had left his fortune to the United States of America to be used "for an increase and diffusion of knowledge." Smithson had never visited America. No one knows why he left the American people his hefty fortune, which came to him through his mother, an English royal. Nevertheless, his bequest has been successful, probably beyond his wildest dreams.

In the United States, the English coins were melted down into about half a million dollars' worth of gold. After nearly a decade of debate—states' rights senators opposed acceptance of the bequest because the institution would increase the power and prestige of the central government—the money gave birth to the Smithsonian Institution, today the world's largest research institution with about five hundred staff scientists and another five hundred or so scientific fellows on temporary assignments.

Among those scientists is Clyde Roper, a man whose
obsession with the giant squid has made him the model for
the main character of several novels and who has been featured,
like Bill Gilly, in numerous television documentaries. To find
out why Roper pursued the giant squid, I visited him in the
office where he's worked since 1966. Walking to his lab through
the Smithsonian's maze of windowless hallways was like visiting
ancient, musty catacombs. A faint odor of decayed flesh seemed
to waft through the air.

Roper's own rooms are filled with bits and pieces of squid and
other animals stored in formalin-filled, clear glass specimen jars;
with stacks of files containing his numerous research papers;
and with all the flotsam and jetsam and detritus and sediments
that have accumulated in corners and on top of filing cabinets
and shelves throughout the course of his forty-five-year career
in science. His place could be a museum in its own right.

Science has been good to Roper, but he has certain regrets.
His professional life's compulsion, to find a live giant squid,
has not been achieved. I asked him about the roots of his Ahab-
like fixation.

His bushy eyebrows arched.

"You don't work in cephalopods for very long without realizing
that the big one is out there," he told me in his Downeast New
England accent.

Growing up on the edge of the North Atlantic, Roper had
always known about the giant squid, but his interest changed
to something more compelling after an event that occurred
on Plum Island, along the northern Massachusetts coastline,
quite near where Roper grew up and only a few miles from
where Rob Yeomans teaches high school marine biology.

Roper told me the abbreviated version of his first giant squid
encounter. Curious, I did a little research of my own.

· : · : · : ·

On a frigid winter morning in February 1980, thirty-six-year-
old Steve Atherton, a family friend of Yeomans, woke with mixed

feelings about the Nor'easter raging outside his Newburyport cottage. The moisture was needed, but the bone-chilling wind blowing in off the North Atlantic would make for a nasty beach run. As Atherton drank his early morning brew, he briefly toyed with the idea of staying warm inside with his wife and another cup of coffee. Then habit took over. His daily run—spring, summer, fall, and winter, no matter what the weather—was a matter of pride. Atherton opened the door and trotted down the desolate beach.

Plum Island, a nine-mile-long, six-thousand-year-old barrier of sand and rock that protects the mainland from the open North Atlantic, remains mostly in its natural state. When Atherton ran over the sand that frigid, windy morning he saw what he thought was a log washed in by the gale. But both the color and the shape were wrong. Finally, only a few yards away, Atherton understood: It was a squid. A huge squid, of a size unlike any he'd ever seen.

It seemed nearly dead, but its eyes were still clear. Its two feeding tentacles were missing and the thing was huge, almost beyond imagination. But it was mostly the eyes that he would remember. They were as large as dinner plates. They still seemed, even in that deathly state, to have an eerie ability to follow prey with an intense and unwavering focus.

Atherton thought he was alone on the beach that morning, but from a distance another person had seen the same animal— from a patrol vehicle.

I hope that isn't a whale, Bill Papoulias thought to himself. *Maybe it will wash away.*

Papoulias, the local federal Fish and Wildlife Service officer, didn't like whales on his beach. Whales were a pain in the derriere, particularly dead whales or, even worse, dying whales. There'd be paperwork, burial detail, and slews of annoying press calls and stressed-out animal activists. A dead or dying whale was not what he needed, not this early in the morning and not on this nasty day.

Maybe, Papoulias hoped, whatever was out there would go back where it came from. He continued his patrol but on the return trip saw that the thing had washed up even farther. It was now solidly beached above the wrack line. It wasn't a whale at all, but a massive squid. This was good from the point of view of paperwork, but not so good when it came to figuring out what to do with all that dead flesh.

Papoulias figured that since it was so large, he ought to report the carcass. But who would you report a squid to? He tried the hotline for Boston's New England Aquarium, but the weekend operator was blasé.

"If it's not a marine mammal, we don't handle the problem," she answered.

Next Papoulias called Bill Coltin, the photographer for the local newspaper. Coltin was intrigued. Papoulias picked up the photographer and graduate student Barney Schlinger and took them out to look at the animal. When Schlinger saw the huge carcass, he was awestruck. He and Papoulias started to examine the squid while Coltin took photos nonstop. No one yet knew exactly what kind of squid it was, but they all knew it was a good story.

When the photos went out over the newswire, it was finally identified as an *Architeuthis*—one of the few nearly intact specimens recovered up to that time.

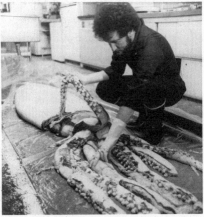

The Plum Island squid

Papoulias just wanted it off his beach.

One man's headache is another man's treasure. The aquarium operator, it turned out, had forwarded the message. Aquarium biologist Greg Early, annoyed by the request to drive an hour from Boston to Newburyport to respond to a squid, called Papoulias.

"Put it in a bucket and we'll get up there and pick it up," Early said.

"Uh, I don't think it will fit," Papoulias answered.

A bigger-than-bucket-size squid . . . now Early was interested. Who knew what this might be? Plum Island was a place where strange things from the North Atlantic often washed up. He and an assistant drove up to take a look.

When the pair saw the animal, Early decided he had to have it. He dug it out of the frozen sand, then recruited a crew to carry it to his truck. It took ten men and a stretcher.

At the aquarium, the team preserved the specimen, but no one knew exactly what to do with it. From mantle tip to arm tips, the thing was 30 feet long. Harvard's Museum of Comparative Zoology considered providing a final resting place, but museum officials nixed that idea. They worried that the floors in the classic old museum building wouldn't be strong enough to hold it. For a while it sat in the front entry area of the New England Aquarium, but staff complained about the aroma.

Clyde Roper, thrilled that such a rare specimen had washed up on a beach not far from his childhood home, drove all the way up from Washington, D.C., to take a look. Roper decided the carcass belonged in the Smithsonian, where there was plenty of room to display it, where the floors were strong enough to hold it, and where thousands of schoolkids every day could have a look at this real-life sea monster.

He squeezed the squid into a coffin, the only container he could find that was large enough, and drove it back to D.C. For years, the squid rested honorably in the institution's Museum of Natural History rotunda beside the 13-foot-tall African bull elephant.

Roper claims that, after the Hope Diamond, the giant squid was the museum's second most popular exhibit. Then the Women's Christmas Committee decided to have their Christmas party in the rotunda. Not willing to host the smelly squid at their gala gathering, the women decreed: The corroded cadaver had to go. For a while, it sat in the basement. Barney Schlinger, with Papoulias the day the animal washed up on Plum Island, ended up heading a UCLA research lab. When he had to travel to D.C. for business reasons, he used to sneak in to visit it hidden away among the dusty old walruses. Today, replaced in the Smithsonian by newer and better preserved specimens, the Plum Island giant squid sits in a glass sarcophagus in the Georgia Aquarium by a display of live whales.

· : · : · : · : ·

Getting his very own dead *Architeuthis* only whetted Roper's research appetite. He yearned to know more about what the animal was like when it was alive. Was it really the dangerous fiend that a thousand years of sea legends claimed it was? Piccot's dangerous adventure seems to have been a fluke, since no one has seriously reported such an event since then, but why had it happened at all?

The only way to answer some of these questions, Roper determined, was to film a live specimen in its deepwater habitat. To find the giant squid, Roper first searched for sperm whales, whose principle diet is squid of many different species. Sperm whales often dive thousands of feet below the water's surface in search of squid—giant, colossal, Humboldt, and otherwise.

Therefore, Roper reasoned, where there were sperm whales, there would be squid. He and other scientists had analyzed the stomach contents of many dead sperm whales over the years and concluded that one whale may eat as many as forty-thousand squid a week. The species of squid eaten by the whale can be determined by looking at the squid beaks, which do not get digested. By separating out giant squid beaks from the other squid beaks, Roper estimated that a sperm whale might eat one or two *Architeuthis* a week.

"It's fuzzy math, because we don't really have much data," he told me, "but a sperm whale might eat between fifty to a hundred giant squid a year." He suspects that sperm whales are eating only a "minuscule" fraction of the number of *Architeuthis* in the ocean depths. "There's a lot of giant squid out there, but we don't see them, because we don't live where they live. Sperm whales and giant squid are neighbors who share the same feeding ground." When a giant squid dies several thousand feet down, chances are good that the body will be consumed by other predators before any of its parts have a chance to float to the surface. Consequently, Roper suggests that *Architeuthis*, far from being rare, may be a fairly common deep-sea species.

In 1996 Roper and Greg Marshall of National Geographic put an inflatable kayak in Atlantic waters around the Azores islands and paddled over to several female whales. I asked Roper if it was dangerous to interact with such a large animal while in such a tiny vessel.

"Their reactions were variable, but never violent or aggressive," Roper said. "You have to approach them slowly and carefully and from behind." Sperm whales, when hunting, surface between dives for only about twelve minutes at a time. In that short period, Roper and Marshall managed to attach submersible cameras to the heads of two whales. After about an hour, the cameras popped off and the team retrieved them to look at the footage. They heard a lot of whale vocalizations and got images of a variety of deep-sea life. Fascinating. No giant squid, though.

In 1997, Roper and his colleagues took more high-tech equipment, including an underwater roving vehicle, to a sperm whale haunt off the New Zealand coast. Again, no luck. There were plenty of sperm whales, but the expedition did not catch an image of a giant squid. In one final attempt, in 1999 he and his colleagues went to Kaikoura Canyon, a deep-sea location off New Zealand favored by sperm whales. Six specimens of giant squid had been brought up by the deep-sea fishing fleet, so the canyon looked like a good bet. But again, no luck. Whales, yes. Giant squid, no.

"I'd go there again if I had the funding," Roper told me wistfully. One of his big issues is the lack of funding for ocean research, something he feels is both unwise and unjust. "Why aren't we spending billions studying our oceans?" he asked me.

"We know more about the moon's behind than we do about the ocean's bottom."

·   ·   :   ·   :   ·   :   ·   :·

Roper wasn't the only Ahab in search of a live giant squid. Finally, in 2004 and 2005, Japanese scientist Tsunemi Kubodera succeeded in his own quest. Also using sperm whales as guides, Kubodera's research team took the first photos and video of a living Pacific *Architeuthis* swimming in deep waters near the Japanese coast. The first picture was of a live giant squid that the Japanese team had caught and brought up to the surface. It died soon after, but the team was able to get a few photos before the animal's demise. The second opportunity allowed Kubodera and his team to video the giant squid in its deep-sea habitat using its feeding tentacles to try to capture prey. The news flashed around the world: The giant squid had finally been located and filmed in its deep-sea habitat. Roper congratulated his successful colleague with a "job well done."

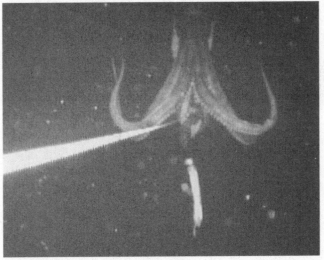

A Kubodera screengrab of a giant squid underwater

For at least the past thousand years, and perhaps even longer, people have debated the question of how dangerous *Architeuthis* might be. If you just look at the size and power of its tentacles and at the teeth on the suckers on the tentacular tips, you'll probably come to a pretty grim conclusion. Many scientists accept the gist of the tale told by Theophilus Piccot—that he and his assistant were aggressively attacked—although they question the information the men provided as to the size and vitality of the animal. Piccot said the animal's body was 60 feet long, but of course, he hadn't stopped to measure. In the popular imagination, the Kraken is both huge and vicious. In the popular novel *Beast*, Peter Benchley's fiendish squid is about 100 feet in length.

In reality, the animals studied recently have been about 40 feet or less, measuring from the tip of the mantle to the tip of the arms. The mantle itself often measures seven feet or less. Piccot described the beak as quite large, but modern studies show that most giant squid beaks fit easily in the palm of a man's hand. It's possible that the animals were larger in Piccot's day— the general consensus is that sea life has decreased in size over the past century—but it's also possible that the fishermen may have amplified certain aspects of their experience.

Some of the exaggeration may be due to the elasticity of the squid's two feeding tentacles, which can sometimes stretch to many times their normal length. When a squid is dead, the feeding tentacles lose their elasticity completely, so that some descriptions may overestimate the true tentacle length of a living animal. The late South African squid expert Martina Roeleveld believed that the tentacle length of a living *Architeuthis* is in fact quite varied. Her own measurements of numerous specimens showed that the tentacle length may be anywhere from 23 percent to an amazing 832 percent of the length of the specimen's mantle.

What is the temperament of a giant squid? Is it laid-back, or is it a rapacious hunter? Does it pass its time suspended in the water column, waiting for unsuspecting prey to drift by, as do some other mid-water species? Does it live alone, with a few other squid, or in large groups, like *Dosidicus gigas*? Many people claim the giant squid is quite aggressive. Piccot described the animal as dangerous, but he had had an unfortunate encounter, which likely influenced his fact-finding. Some suggest that the animal is benign. The truth is probably somewhere in between.

I, for one, wouldn't want to meet one by accident while diving or in a small boat.

Roper, on the other hand, would probably be delighted to do just that. Of course, he's known as somewhat of a fanatic. A YouTube animation from Britain has him answering the question "What do giant squid eat?" with: "Anything they want." At the same time, the animation shows several muscular and menacing giant squid arms pulling the bearded scientist and his little skiff down under the gently roiling sea surface.

He's grinning as he goes down with his ship.

I asked: Would he paddle up behind a surfaced *Architeuthis* as he did with the sperm whales?

"Absolutely," he said. "I wouldn't have any real concern about that, as long as I had a waterproof video camera, maybe with a flotation device for the camera. It would be absolutely thrilling. Can you imagine?

"Can you imagine," he continued, "how absolutely spectacular it would be to swim along with them? That gigantic eyeball as big as your head! I'm not sure I'd look forward to a gigantic embrace. . . ."

Truly a man possessed, I thought to myself. I asked if he discounted the tale of Theophilus Piccot.

He said he'd take it with a grain of salt. No one really knows much about the animal's temperament. About Piccot, Roper said: "A giant squid that appears at the surface is not normal. It's either dead or dying. The tales sound wonderful and exciting, but for somebody like me, you want to deal with the truth. How do you determine truth? Show me the evidence—a confirmable photograph or video, something that really will prove that the animal was there and alive and vigorous."

Nevertheless, in some of his writing, Roper exhibits no Gentle Giant illusions about *Architeuthis*. "Two-thirds of their total length consists of two long, bungee cord–like tentacles with sucker-studded clubs at the ends, used for capturing their prey like a two-tongued toad," he once wrote.

For quite a while, most people believed that the giant squid was the world's largest and most ferocious invertebrate. But recent studies of the colossal squid, which lives in the Southern Ocean that surrounds Antarctica, show that this animal may far outshine the giant squid in these matters. The mantle of the colossal squid, *Mesonychoteuthis hamiltoni*, may be twice the length of the giant squid mantle, and measuring from mantle tip to tentacle tip, the animal may be as long as 60 feet. We don't know for sure, since we haven't recovered enough intact colossal squid specimens to be sure. But we have recovered quite a few colossal squid beaks. French scientist Yves Cherel and Canadian Keith A. Hobson analyzed the chemistry of those beaks, as well as of giant squid beaks, in order to learn more about the diet of both species. They found that the colossal squid eats much higher up the food chain, like sharks and some of the toothed whales, than does the giant squid.

Nevertheless, the photos and video of the live giant squid taken by Kubodera seem to show a highly focused hunter with powerful feeding tentacles. In a 2005 *Royal Society* paper, Kubodera and coauthor Kyoichi Mori described their September 30, 2004, encounter. The team dropped a line baited with smaller squid to a depth of about 900 meters, or roughly 3,000 feet, into a deepwater canyon near the Ogasawara Islands off the coast of southern Japan. They also dropped a deepwater camera, which caught images in which an *Architeuthis* approaches the bait with open arms. Next, the two tentacles flash out. One, hooked by the squid jig, breaks off after several hours and is brought to the surface.

"*Architeuthis* appears to be a much more active predator than previously suspected, using its elongate feeding tentacles to strike and tangle prey," the two scientists wrote. "The recovered section of tentacle was still functioning, with the large suckers of the tentacle club repeatedly gripping the boat deck and any offered fingers. . . . Giant squid are unique among cephalopods as they can hold the long tentacle shafts together with a series of small suckers and corresponding knobs along their length that enable the shafts to be 'zipped' together. This results in a single shaft bearing a pair of tentacle clubs in clawlike arrangement at the tip."

Reading this, I wondered how many swimmers would be jumping into the water alongside Roper, were he ever fortunate enough to find an *Architeuthis* to swim with.

A Kubodera photo of a giant squid being captured on the surface of the water

· : · : · : ·

Because we have more ships at sea these days, we're recovering more intact specimens, as well as more giant squid bits and pieces. In the summer of 2009, on a cruise a few hours out of Los Angeles, scientists from Scripps found a piece of giant squid. Around the same time, federal scientists pulling a deepwater trawl in a Gulf of Mexico area frequented by sperm whales also hauled up a giant squid. Since a carcass had been found on a Louisiana shoreline in the 1950s, scientists were not taken completely by surprise, but neither was the Gulf of Mexico team expecting the specimen.

On August 24, 2002, bathers at a popular Portuguese swimming beach noticed something strange in the water that turned out to be a dead juvenile giant squid, the first ever reported off this part of the coast of Portugal. Scientists were surprised to find an *Architeuthis* that far south, as many had previously believed that their regular Atlantic Ocean habitat was farther north. Was its presence an indication of a change in ocean ecology? And off the coasts of Japan and New Zealand, the number of recorded specimens seems to have increased in recent decades.

Not all reported sightings turn out to be accurate. A "giant squid" reported by the press in the Caribbean's Cayman Islands in the fall of 2009 turned out to be only a very large squid of about six feet in total length. It wasn't an *Architeuthis* at all, but an *Asperoteuthis acanthoderma,* a species that may be spreading from its former Pacific Ocean habitat. (Or, given our lack of knowledge about life in the ocean, it may have been in the Caribbean all along, but not specifically identified.) Some of the confusion comes from reports in the press. The media frequently calls *Dosidicus gigas* a "giant squid," meaning that it's a very large squid, rather than a squid of the *Architeuthis* group.

The confusion is understandable when you consider that even scientists are sometimes uncertain which squid species is which. At a scientific conference in Portland, Oregon, a federal marine biologist approached Gilly to say that recent deep-sea exploration vehicles had filmed many Humboldt squid swimming in deep-sea canyons along the northwest coast.

"In the canyons?" Gilly asked, surprised. "We'll have to come up and take a look." Humboldts had been caught in the Pacific that far north, but large numbers using the deep-sea canyons implied that they may not have been just passing through.

"Of course," Gilly speculated, "it could be some other large squid species we don't even know about." No one knows if some species of squid are truly becoming more common, or if the animals are only being found more frequently than before because we are able to penetrate the ocean depths with more technology.

If more *Architeuthis* are showing up in the world's oceans, does that mean there are more living now than in past centuries? Or does it just mean that there are more of us out there looking for them? If there are more, should we be afraid? For centuries, people have fantasized about the abilities of all these large squid species—giant, colossal, Humboldt, and others—to the extent that I often wonder if there isn't some kind of vestigial horror of being entwined in all those arms embedded in the evolutionary recesses of our brains. In Peter Nichols's *A Voyage for Madmen*, a nonfiction account of a solo round-the-world race, one of the sailors confuses nighttime phosphorescence in the waves with the eyes of a giant squid, which he tries to kill with a harpoon.

Even Rachel Carson indulges in a tiny bit of fearmongering: "We can imagine," she writes, "the battles that go on, in the darkness of the deep water, between these two huge creatures— the sperm whale with its 70-ton bulk, the squid with a body as long as 30 feet, and writhing, grasping arms extending the total length of the animals to perhaps 50 feet." Carson wasn't intentionally exaggerating the size of the giant squid. During her era, many scientists thought the animals were that large because of the size of the circular squid sucker scars found on sperm whale skin. But Roper speculates in one paper that those scars might be that large because they were made when the whales were smaller, and expanded in diameter as the whale grew larger.

Of course, what people really want to know about *Architeuthis* is how smart it is. The question of squid intelligence in general has piqued the curiosity of marine scientists for several decades. Gilbert Voss, Roper's doctoral adviser, wrote in a 1967 *National Geographic* article that "some squid exhibit behavior bordering on active intelligence."

I asked Roper what he thought about cephalopod intelligence.

His answer was a big question mark with a provocative caveat: "When you look into their eyes, you know there's something there," he said. "But be careful how you use the word 'intelligence.' We use the word but don't try to imply any kind of human characteristics for mollusks or any invertebrates. They do, however, show a great deal of brainpower."

Gilly is less reticent. "They can certainly respond to novel situations in appropriate ways," he said. "I could call that intelligent." He mentioned some BBC footage he'd seen of large numbers of *Dosidicus* swimming together in a highly coordinated way, as though their movements were choreographed. They seemed to be moving in unison, as though they were able to communicate with each other. "They were turning on a dime," he said. "It was really, really beautiful."

· : · : · : · :

On the Monterey Bay research boat, the severed head of a Humboldt squid flew out of Julie Stewart's hands and across the boat deck. The lethal beak slashed a "V" in her thick, protective Grundens. Carrying another squid, she slipped on the slime of *Dosidicus* ink that covered the deck. Her hands were cold. Strands of ink-covered hair escaped from her ponytail and got in her mouth. Her back was tired. The skin on her fingers was lacerated with countless tiny cuts. Marine biology is not a science for the faint of heart.

Stewart at the center of the action

None of this bothered Julie. "You take a shower when you're done," she said. "Your hands still smell like squid for a couple of days. You get squid ink under your fingernails, but it doesn't stay long. That's just part of the job." Then she trailed off and shrugged her shoulders . . . not a problem. Once when she was asked what she liked most about field science, she gushed. When she was asked about what she liked least, she couldn't think of anything. She mentioned that spending time on a boat in rough water makes a lot of people seasick: "But I don't get seasick."

By 7 p.m., Julie had finished tagging. The shoal of Humboldts was still churning up the water all around the boat, so the team decided to harvest squid stomachs. By recovering the contents of the stomachs for later lab study, the team would move a step closer to answering another pressing question: What do Humboldt eat while they're in Monterey Bay? Some researchers had found a correlation between declining numbers of hake—a Pacific whitefish that often ends up in fish sticks and other assorted fish products—and the presence of large numbers of Humboldts. The scientists wanted to harvest the Humboldt stomachs to confirm those findings.

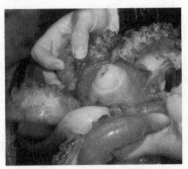

Stewart uncovers the squid's beak

Julie cut the brain stem of each squid with a knife, right below the head. She didn't seem to mind. "Field biology itself selects for human researchers with particularly rare characteristics," molecular biologist Sean Carroll once wrote. In Julie's case, that seems to be true. She spent the rest of the evening working on another fifty or so *Dosidicus* specimens, measuring their lengths, severing their brain stems, and cutting out their stomachs.

When the researchers hauled the squid on board, the animals continued to flash red and white colors by using their chromatophores, but when Julie made her cut, the squid's body turned from red to white in milliseconds. The tentacles, however, with their own sets of nerve cells, remained red. "The nerves from the brain to the tentacles don't go through the part of the brain she cut," Gilly explained later. "When you cut behind the head you're severing the anterior and posterior parts of the animal from nervous control. But not the arms."

LUMINOUS SEAS

It's almost like their skin's covered in television screens.

—MARK NORMAN, AUSTRALIAN CEPHALOPOD SCIENTIST

Australia's Great Barrier Reef is the planet's largest living superorganism, a 1,600-mile, coral-based amalgam of thousands of living species that depend on each other for survival. Among these are more than four thousand mollusk species, including hundreds of species of cephalopods. Some of them have been identified, but many have not. Finding out what's there is an immense undertaking.

Among those trying to accomplish this goal is Australian researcher Mark Norman, who has identified more than 150 new mollusk species. Also an expert underwater photographer and videographer, Norman has dazzled people around the world with his work, which shows the profound variety of colors, shapes, and lights that can be used by cuttlefish, octopuses, and squid. His work has brought the once-arcane subject of cephalopods out of the science labs and into the public eye. It was Norman's video of a coconut-shell-carrying octopus that wound up on YouTube and went viral, enjoying more than a million views in only a few weeks. Knowing I would be interested, at least twenty different people sent it to me.

Later Norman and his colleagues published a paper on the phenomenon, calling it "tool use," since the octopus seemed to be carrying the shell around for future use as a shelter. In the video, the octopus carries the shell directly under its body by gripping it with some of its arm suckers. The animal then moves across the seafloor by walking on some of its arms, a difficult and burdensome activity that the scientists dubbed "stilt-walking."

Another Norman video that shows a mimic octopus that could change shape and color, appear and disappear before viewers' eyes in only milliseconds, has been seen by more than

500,000 people. The ocean is a riot of color and light, and no group of animals makes better use of the available palettes and shadings and light shows than do the cephalopods. Some, like cuttlefish, use rich repertoires of colors on their skin, lots of letters in their alphabets, to communicate, causing Norman to equate their skin with television screens. But cephalopods use much more than color when they call upon their quick-change, now-you-see-them, now-you-don't artistry.

Most of the ocean, down below the surface, is dark—darker than the darkest night. Strong sunlight penetrates only to the top several hundred feet. Most light has disappeared by 600 feet down. Below that layer is a mysterious netherworld of half-light, the twilight zone. Several thousand feet below the surface, even that tiny bit of light has disappeared. Instead, this world is filled with *pinpoints* of light—light not from the sun, but from whatever life is down there. If you could swim in this world (you can't because the pressure would kill you), it would seem as though you were a spaceship swimming in a light-filled ocean of twinkling stars.

In many ways, the sea is more celestial than earthly. These pinpoints of light—bioluminescence—come from the life that thrives in the water. In the sea, the ability of animals to make their own light is more the rule than the exception. Probably about 90 percent (or more) of the ocean's species can produce some kind of glow. Sea life uses this light in a wide variety of ways—as illumination in the dark sea depths, as a way to lure prey, as a way to confuse predators and gain time to escape, as a communication strategy, and even as a sexy statement to other members of its own species.

Twenty-five hundred years ago, Aristotle, arguably the world's first marine biologist, wrote about the strange "cold" lights in the ocean, and we have been attracted by the light show ever since. People have found ingenious ways to use this biological phenomenon. Some indigenous tribes captured animals that glowed in the sea and used them as "flashlights" to light up nighttime paths through the forest. In Paris in 1900, one inventor filled a glass bulb with a glowing bacterium that's quite common in the sea,

Vibrio fischeri, and used it to light the interior of a room with a strange, otherworldly glow. We ourselves like to mimic this phenomenon: Around the holidays, when the northern half of the planet is at its darkest, we string tiny blinking lights on our houses and down our streets to comfort us until the sun begins to return. Perhaps our fascination with tiny lights twinkling in the dark is vestigial.

That the sea should be filled with light isn't surprising, since sensitivity to light was one of life's earliest characteristics. A simple response to light, angling toward the rays of the sun, is elemental. Even plants do it. The ability seems to have been built into the structure of simple one-celled animals. In recent decades, researchers have discovered one particular gene (sometimes called "eyeless" because without it there would be no eye) that's present in such a wide span of animal life— houseflies, mice, simple wormlike sea animals, cephalopods, and us—that they believe this gene has existed since evolution's early days. The eyes in the animals with this gene are not all alike, nor are they all equally efficient. But the presence of very similar eyes in so many species points toward one of science's most momentous breakthroughs: understanding the universal interweaving of all life. We know about evolution from Darwin's point of view, but in recent years scientists have come to tease out the details of the fundamental basis of genes and of DNA. And it turns out that, in this fundamental sense, there's nothing new under the sun: The genes that helped make early proto-eyes are also present in our own bodies, in modified form, and they're the reason why you can see well enough to read this book.

. : . : . : .

The ocean's light shows have always attracted our attention, but no one until very recently realized that this radiance could create a revolution in medical research that would save human lives, winning scientists a Nobel Prize in the process. In the 1960s, a young Japanese scientist, Osamu Shimomura, working at Friday Harbor Laboratories near Seattle, Washington, was curious about why a certain kind of jellyfish swimming in nearby waters gave off an unusual green glow. He found that the glow was caused by a specific and unusual protein, which he identified.

Shimomura published his finding, which was largely ignored for several decades. Then another researcher, Douglas Prasher, learned how to manufacture very large amounts of this protein in his laboratory. For the first time, it was widely available to other scientists. Then two other researchers figured out how to use that strange glow to light up proteins inside an important type of brain cell, the neuron. The green light allowed neuroscientists for the first time to watch the inner workings of the neuron on a very basic level, on the level of molecules at work inside the cell.

By "tagging" or attaching the protein that lit up green to the other molecules at work inside the cell, researchers could watch those now-glowing molecules do their day-to-day jobs, keeping the neuron functioning. Today, hundreds of researchers are using this breakthrough to study what goes wrong in neurons when people develop diseases like Alzheimer's or Huntington's.

Fascinated by the green light—called "green fluorescent protein"—Yale University neuroscientist Vincent Pieribone wrote a book explaining the wide-ranging importance of Shimomura's serendipitous and seemingly (at first) unimportant discovery. To Pieribone, scientists are voyeurs who try to spy on the long list of "individuals" at work keeping things shipshape inside cells. "Proteins are the workers of the body," he told me. "They have these little individual personalities, like the guys who pick up the garbage, the guys who drive the trucks. We want to know as much as we can about the lives of these proteins, these amazing little guys. What's their lifestyle like? Do they change their personalities?"

The key to this breakthrough is that the green fluorescent protein can illuminate these various individual proteins without disturbing their activities. It's kind of like asking some members of a crowd to carry light sticks so observers can follow them in the midst of the mass and see where they go and what they do when they get there. "This has transformed science and medicine," Pieribone continued. "Now, we can look inside a brain cell with molecular specificity." Formerly, researchers had to destroy cells in order to study them, so that a scientist might be able to look at an individual protein frozen in time but wouldn't be able to see what the protein's job was. Using the new tagging technology from the jellyfish, Pieribone and hundreds of others now watch the many protein-workers in real time as they move things around in cells, as they clean up messes, or as they repair damage.

I asked Pieribone if anyone expected a jellyfish to help find a cure for neurological diseases back in the 1960s, when the phenomenon was first discovered and reported to the scientific community.

"No," he answered. The original scientist had just been indulging his own curious nature: Why did that particular species of jellyfish display a particularly unusual green glow?

"Very interesting findings come from very strange places," Pieribone concluded.

The breakthrough, he added, took a series of scientists several decades to discover, and comes under the category of "obscure studies that made huge contributions to the world."

Some squid research exemplifies this pleasantly serendipitous phenomenon quite well, he said, adding that it would be an enormous mistake not to study the animals: "It's been exhilarating to learn how bizarre the world is under the ocean. I happen to have a huge fascination with squid. When I see them when I'm diving, I try to get as close to them as possible. You recognize that they have a certain level of intelligence. They and other animals in the ocean can provide fantastic tools that don't exist in our own world. We can capture these tools from other animals and have amazing libraries of solutions."

"Amazing libraries of solutions . . ." I thought it was an interesting concept, well put. Jellyfish, of course, have contributed green fluorescent protein to our modern medical toolkit.

Other sea life has been equally helpful. Only recently, Japanese researchers developed a very promising treatment for women suffering from advanced breast cancer. The new medication derives from DNA taken from a species of sponge. There are plenty of other examples, and, Pieribone says, we've only seen the tip of the iceberg.

. : . : . : . :

Cephalopods are the masters when it comes to creating light shows under the sea. They can use light to disguise themselves, to hide from predators, to lure prey into their waiting arms, or maybe, sometimes, just to see better. Their strategies for light-manipulation seem to be boundless. Julie's Humboldt squid was covered with bioluminescent photophores—small, light-emitting packets embedded all over the animal's skin. Another squid, *Heteroteuthis dispar*, a tiny deep-sea squid, is nicknamed "fire shooter" because, instead of shooting out a cloud of ink to confuse predators, as do many cephalopods, it shoots out a cloud of light. The sudden, unexpected light distracts the predators, buying the squid enough time to jet away. Perhaps the combination of genes that triggers the fire shooter's remarkable behavior could one day be used to help us in some way.

The colossal squid, which probably lives about 3,000 feet below the ocean's surface, has eyes that can grow to nearly a foot in diameter. Attached to the back of the eyes are bioluminescent "headlights" that provide extra light. The "headlights" help the animal attack its prey in the darkness of the deep ocean by providing enough light for the animal to judge the distance from its eyes to the attack point.

The truly weird, eight-arms-but-no-feeding-tentacles squid *Taningia danae* specializes in shock and awe. It has large bioluminescent organs on its arm tips that flash just when the animal attacks. Scientists think the light may perform two functions: confusing the prey and establishing the correct distance of attack.

In the mid-level layer of the ocean where some sunlight penetrates, squid have developed a strategy called "counter-shading" to protect themselves from predators. Clyde Roper and other scientists have found that some species of squid can produce light on the ventral, or lower, surfaces of their bodies. The animal is able to make the level of light coming from its lower-surface photophores match the level of sunlight in the water above so that the animal disappears. Predators looking up from deeper in the water see a dappling of light coming from the rays of the sun. The animal's belly produces the same dappling, so the predator sees nothing to break the pattern of the sunlight. Roper and his colleagues also learned that the animals can change the amount of light coming from these lower-surface photophores within a matter of seconds. When the scientists changed the level of light coming down from above an experimental animal, within half a minute or so the squid adjusted the amount of light coming from its lower-surface photophores, continuing to match the light from above.

Meanwhile, despite the amount of light coming from above, the dorsal, or top, surface of the animal remained dark. Predators swimming above, looking down into the ocean depths, saw only black.

· : · : · : · :

The light of bioluminescence comes from chemical reactions. Squid can produce light in different ways, but often call upon friendly bacteria to do the job for them. From the point of view of the squid, the bacteria living inside its body are a worthwhile investment because the payoff of added light helps the squid thrive. From the perspective of the bacteria, the squid provides room and board. It seems to be a happy marriage of satisfied coequals.

"It's a deep conversation between two partners," Margaret McFall-Ngai explained to an audience in a Marine Biological Laboratory auditorium one hot July evening in 2010. The weather was sweltering and the lecture hall had no air conditioning, but her remarkable talk, "Waging Peace: Diplomatic Relations in Animal-Bacterial Symbioses," easily held the attention of several hundred scientists. This was cutting-edge stuff.

She was explaining that bacteria living in host animals aren't just hitchhikers looking for a free ride. They give something back to the host, although the relationship can be pretty complex. For example, in humans, a certain type of bacteria that's essential to intestinal health and, strangely, the third wave of the sleep cycle, can also cause whooping cough and gonorrhea.

Scientists suspect that the bargain between bacteria and host has been present from the earliest days of evolution, since bacteria in one form or another have been around for well over a billion years. In fact, from the point of view of bacteria, the purpose of evolution might be simply to provide the bacteria with a wider array of housing options.

Since both humans and squid act as host animals to bacteria, findings from studying squid and bacteria can help further human medical research. The work done by McFall-Ngai and her colleague Edward Ruby has even helped doctors understand why, whenever possible, human babies should be delivered naturally, rather than by cesarean section.

"We are not the single individuals that we think we are," McFall-Ngai told me. "You have all these bacteria living with you that are required for your health. I study squid, but the major focus of my lab is to inform the biomedical community as to how bacteria form these persistent relationships with animal cells."

McFall-Ngai has shown this by studying the Hawaiian bobtail squid, *Euprymna scolopes*—"the couch potato of the squid family." Temperamentally, the bobtail squid is a rather laid-back little animal. Full-grown, this unobtrusive fellow is small enough to fit in the human hand. To me, its behavior more closely resembles that of cuttlefish than of the Humboldt. Because of this, it's easier to keep in a research laboratory than a more reactive squid.

Unlike most squid, the bobtail squid spends much of its daylight hours buried in the sand. When the sun goes down, the bobtail comes up. It begins hunting for prey. This is a good protection strategy, but the vulnerable little animal, lacking the ability to move quickly, needs help in order to survive. Although it hunts at night, moonlight and starlight make it visible in the shallow waters it favors, making it both vulnerable to predators and avoidable by prey.

To solve the problem, it camouflages itself by acquiring bacteria that give off light. This sounds counterintuitive, but it turns out that this teamwork between squid and bacteria pays off: The light from the shining bacteria carried by the squid helps the host squid better blend in with the light from the moon and stars above.

The squid has even evolved a special place to house the bacteria—a structure scientists call a "light organ" located inside the mantle. But the bacteria don't just set up house and begin to party. When the first batch of bacteria arrive in their new digs, they have work to do. The light organ has been partially prepared for the bacteria's arrival, but the final touches won't occur until the newly arrived bacteria get things started. It's as though they've arrived in a new house but only the frame is up. The bacteria themselves have to do the finish work.

The bobtail squid and its bacteria must begin a kind of holding of hands on a cellular level. This brings the squid's light-organ cells to maturation. Unless these squid cells change, the animal's light organ will not function, the squid will not be protected, and it will likely become somebody else's dinner.

Still, the squid is a hard taskmaster: Once the squid has the bacteria, and the bacteria are living and multiplying in the light organ, the squid doesn't keep the same specific living bacteria from one day to the next. Each morning, at just about dawn, when it is done hunting and is preparing for its rest period snuggled in the sand, the bobtail squid expels almost all the bacteria from the day before. Over the next hours, the bacteria remaining in the squid's light organ multiply and multiply until the necessary number are present when the squid goes out to hunt again that night.

But here's the really intriguing point: The bobtail squid doesn't acquire just any bacteria. Only one particular kind will do, the luminous and ubiquitous *Vibrio fischeri,* the same species used by the French inventor in 1900 to make lights for people.

If the wrong kind of bacteria try to enter the squid, the squid rejects them. The squid has no choice in this matter. Its very health and survival depends on its finding and nurturing the right bacteria. Hardworking *Vibrio fischeri* are so essential to the bobtail squid that the animal constantly monitors the light productivity of the bacteria. If the little organisms are not doing their job well enough, the squid evicts them.

The squid is not born with these bacteria. Just as a human baby develops in the uterus free of any and all bacteria, the bobtail squid develops in the egg in a sterile environment. When it hatches, one of its absolutely essential tasks is to find and nurture the rod-shaped *Vibrio*. The human baby must also acquire the correct bacteria.

The parallels between squid and human go even further. When the squid finds and nurtures the right bacteria, the bacteria interact with the developing animal and aid the animal in its ongoing maturation. The first batch of *Vibrio* taken in interacts with the squid's cells so that both squid cells and *Vibrio* develop in tandem. Because of similar basic cell interactions, human babies also need the right bacteria to interact with the newborn's stomach and intestinal cells. For example, we need specific bacteria in our gut to manufacture vitamin K, important for blood coagulation, and vitamin B_{12}.

Like the squid, the human baby develops in the uterus in a sterile environment. But unlike the squid, which must go out and hunt for the bacteria after emerging from the egg, the human baby receives the first dose of necessary bacteria in the mother's birth canal. Without the process of a natural birth, the baby may not encounter these bacterial helpers at the appropriate time, and some of the infant's post-birth development may be at risk.

"During passage through the birth canal," explains McFall-Ngai, "we begin to pick up our bacterial partners, which are essential for our health. The squid research, which is far easier than research on mammals, investigates the 'molecular language' that occurs between host and symbiont." Some researchers connect the increase in human bowel disease in the developed world with the increase in C-sections. "We are 90 percent bacteria," McFall-Ngai explained. In mathematical terms, humans have 10^{14} bacterial cells, but only 10^{13} human cells. That is, if you're just counting cells, there are a lot more of *them* than of *us*.

Bobtail squid are aiding research in the field of human medicine in another way, too. After the squid acquires the right kind of bacteria, the bacteria do not bioluminesce immediately. Their light does not shine until they have achieved high enough population levels. When the critical level has been reached, the glow begins. Other researchers are studying this phenomenon, called "quorum sensing," in order to develop a new and better kind of antibiotic for human use. If scientists can better understand the interaction of the bobtail squid and *Vibrio fischeri*, and understand how *Vibrio* bacteria communicate with each other in the squid's light organ, they may be able to find a way to interrupt that communication. By preventing the communication, they may be able to prevent the bacteria from reproducing. And this ability might eventually be the foundation for the development of a whole new line of antibiotics, of medications that don't destroy a bacterial infection but instead prevent the infection from overwhelming the host organism in the first place. The key to developing this new line of drugs, researchers say, is understanding the fundamental relationship between the squid and the bacteria throughout the squid's lifetime.

· : · : · : · :

One of the biggest problems cephalopods face is how to live safely in a 3-D world. When you imagine swimming in the deep ocean, you have to rethink human-oriented concepts of "up" and "down." As rather large surface animals who live on the continental crust, we usually need only be aware of animals living on the same plane that we do: Will we be attacked by a lion? Trampled by an elephant? Usually, "up" and "down" are not words that hold terror for us. We don't fear giant birds swooping down from above to scoop us up and carry us away, and we don't fear giant worms bursting out of the earth's crust to grab us and drag us underground. For the most part, we only need to be aware of enemies that, like us, are firmly rooted to life atop the soil.

But surviving in the ocean is more complex. An animal living in the sea needs to have the responses and defenses of a fighter pilot. The enemy can come from anywhere, from the left or from the right, but also from above or from below. It's a three-dimensional world down there. Skeleton-free cephalopods are particularly at risk, since predators don't need to worry about the bones. "The creatures are really just rump steaks swimming around," Australian Mark Norman once quipped. They need special protection.

In response, the animals have evolved an impressive tool kit of tricks. The bathyscaphoid squid, named in honor of a self-powered sea exploration vehicle that was developed after Beebe's bathysphere, comprise a family of squid that spends its early life, when it is most vulnerable and most likely to turn into someone else's dinner, at the ocean's surface, where there are plenty of small tidbits for a tiny animal to eat. As bathyscaphoid squid develop, they descend deeper and deeper into the water. These squid have evolved a body that's translucent and almost completely invisible. At the top level of the ocean, the water is rich with nutrients. It's easy for the squid, as predators, to find food. Unfortunately, it is also easy in the sunlight to become prey to other predators. But with a body that's almost transparent, these young squid are ghostlike, nearly invisible. Being nearly invisible when tiny is quite convenient. The young squid at the sea surface can easily sneak up on its even tinier prey without being noticed. A prey animal might perceive what seems to be a twinkle of sunlight at the sea surface, only to find itself enveloped in a mass of squid arms and tentacles.

Locating your enemy in the ocean is a 24/7 task. Color and luminosity are both armor and weaponry. Many animals developed the ability to change shape and color to blend in with their surroundings. Some fish can do this, as can some frogs and, of course, chameleons, but no group of animals is as sophisticated in this strategy as are the cephalopods. When we watch these animals zip through a myriad of psychedelic displays in only seconds, we stare, transfixed. But the basic organization of this magic show is simpler than you might think: It's done with three layers of three different types of cells near the skin surface—a layer of chromatophores, a layer of iridophores, and a layer of leucophores.

The top layer of cells, the chromatophores, contains the colors yellow, red, black, or brown. The colors present are species-dependent. The color in a chromatophore cell sits near the cell's center in a tight little ball with a highly elastic cover. When the muscles controlling the chromatophore are at rest, this ball of color is covered over and can't be seen. When a chromatophore is showing, what you're seeing is this little ball, stretched out into a disk roughly seven times the diameter of the at-rest ball.

To operate properly, one chromatophore cell has a number of support cells, including muscle cells and nerve cells. The arrangement is cunningly elaborate. Anywhere from four to twenty-four muscle cells might attach to only one chromatophore. When these muscles contract, pulling on the chromatophore cell, the elastic sac is stretched out, revealing the color inside. When the muscles relax, the ball returns to normal size and the color disappears.

There's a simple way to envision this: Imagine a small, circular sheet of red paper. Crumple it into a tiny, tight ball. The color red is now only a pinpoint. Using your hands—and the hands of up to eleven other people if they're around to simulate the twenty-four muscle cells—stretch the paper out so that it's flattened to its full size. Then crinkle the paper into a tiny ball again. Do that umpteen times a second to simulate flashing. On an infinitely smaller scale, that's how a cephalopod operates one individual chromatophore.

This is enormously elaborate engineering requiring a considerable amount of coordination and support. The muscles surrounding the color-containing cells are controlled by nerves that interact with other nerves. Some scientists think that this complicated system requiring massive amounts of computing power may be one explanation for cephalopod intelligence.

Just below the layer of chromatophores is another layer of cells, the iridophores. This layer of cells shows a different array of colors—metallic blues, greens, and golds. The iridophores do not open and close. Instead, they reflect light. They are sometimes used to camouflage an animal's organs, such as eyes, by shimmering and drawing attention away from the organ. Some scientists have studied this strategy of distraction-by-light-show as a way to improve camouflage for soldiers on the battlefield.

Underneath this layer is the final layer, a layer of leucophores, flattened cells that passively reflect the color of background light, increasing the animal's camouflage.

When I first watched cephalopods showing off their artistic genius, some of their techniques seemed familiar. I knew I had seen this use of color and light somewhere else. Then I remembered Claude Monet's many paintings of water lilies, of haystacks, and of a cathedral at Rouen. Monet painted the same scene many times, but each painting is different because the master could so expertly show the differences created by only slight shifts of light.

Cephalopods are the original Impressionists. I often wonder if the French painters didn't quietly study the cephalopods' techniques. Both the Impressionists' and the cephalopods' light shows provide the illusion of great depth by using luminosity—the reflection of light. Both skillfully use thousands of points of light and color to trick the observer.

But not all cephalopods enjoy equal artistic talent. Cuttlefish, which live nearer the ocean's surface where light still penetrates, are outstanding in their Impressionistic skills. Humboldts, on the other hand, are quite limited. With their highly honed predatory abilities and their large size, they don't need to devote so much energy to disguising themselves. Moreover, since so much of their lives is spent in dark ocean depths, all the Humboldt needs is red chromatophores, which allow it to disappear quickly.

The giant Pacific octopus, probably the largest of the octopuses, also has red chromatophores. Its size and its ability to hide in crevasses lessens its need for color disguises. Since the production of so many dramatic color options is energy-intensive, it's not surprising that the ability disappears in those species that have better survival strategies.

Some octopus species not lucky enough to be large do have entrancing color choices. The tiny blue-ringed octopus of Australia has lots of color options from which to choose and also has various clever camouflaging strategies. It may vanish in plain sight by taking on different colors and forms, but when it's challenged, it quickly rematerializes and shows startling, almost repulsive blue rings. The unusual tone warns predators of its deadly poison. Most octopuses are venomous (the poison is delivered via the saliva), but the toxins are rarely harmful to humans. In fact, medical researchers are studying some of the proteins in octopus toxins in the hopes of improving current cancer treatments. But a bite from the blue-ringed octopus can kill a human by paralyzing muscles and making it impossible for the victim to breathe.

Most cuttlefish are not that poisonous. Instead, they rely on their artistic expertise, which researchers suspect is so finely honed because cuttlefish are such easy prey. Cuttlefish, apparently for eons, have made choice delicacies for all kinds of predators. Dolphins find cuttlefish delicious, but they like neither the cuttlebone nor cuttlefish ink, which upsets the dolphin's digestive system. Australian scientists recently found that dolphins adept at "butchering" cuttlefish gather in the Upper Spencer Gulf. They gorge on an annual holiday treat— the massive die-off of thousands of cuttlefish that have finished mating. The dolphins herd the dying cuttlefish onto sand plains and kill them. Then the dolphins cleverly get rid of both the cuttlebone and the ink. Once the cuttlefish is properly prepared, the dolphins eat the meal.

Unlike other species, cuttlefish cannot escape predators by swimming deeper, because the cuttlebone will disintegrate under pressure. Instead, cuttlefish hide by pretending to be something else. Roger Hanlon, a cephalopod researcher at Woods Hole's Marine Biological Laboratory, is studying cephalopod camouflage abilities that may have military applications, as is the Air Force Research Laboratory in Dayton, Ohio. Recently, the Department of Defense awarded the MBL scientist $1.2 million for a study of "Proteinaceous Light Diffusers and Dynamic 3-D Skin Texture in Cephalopods." The Ohio lab is studying some of the proteins involved in cephalopod camouflage to see if some of those proteins might somehow be used to help soldiers become less visible on the battlefield.

When I visited the Hanlon lab, I watched as a cuttlefish rested on the bottom of a tank covered in small bits of sandy-colored gravel. The animal's body and some of its arms took on the color of the sand, but two of its arms waved languidly in the water. Next to the animal was a plant with dark-colored fronds. The cuttlefish's two waving arms took on the same dark color and looked just like the plant fronds.

The effect was spectral. It was also dismaying to realize that my own eyes were so easily fooled. We humans are supposed to enjoy the advantage of terrific eyesight and a brain smart enough to perceive what it is we're looking at. Now it turns out that our eyes are not quite as special as we once believed.

In fact, some scientists suggest, the cephalopod eye may actually be superior in many ways to our own.

· : · : · : ·

The *urbilateria,* the hypothetical last common ancestor shared by humans and cephalopods, probably had the ability to respond to light in some way. These early "eyes" were probably very simple—maybe little more than small indented cups in the animal's outer surface. The cups wouldn't have "seen" in the sense that

Rob Yeomans holds out some Humboldt eyes

93

we understand the meaning of the word. Instead, the indentations would have contained compounds capable of responding to light, an ability that would have provided some kind of evolutionary advantage. When other species evolved a method of counteracting this simple eye, more complicated eyes would have appeared. Eventually, the complicated camera eye—the eye we possess—evolved. Some scientists believe that the development of our sophisticated eye was the result of an ongoing evolutionary arms race.

We consider our eye, the camera eye with its marvelous lens and cornea, to be the best available. In fact, when Charles Darwin thought about the human eye, he wrote that it caused him to doubt his own theory. How could natural selection, with its seeming randomness, account for such a complicated and finely tuned organ?

When scientists began studying the cephalopod eye, they found that it, too, was quite complicated. Despite our evolutionary distance from each other, the cephalopod eye and the human eye are strikingly similar in that we both possess nature's most complex style of eye, the camera eye. This similarity turned out to be one clue that helped scientists better understand the eye's evolution, and evolution in general.

But there are important differences. For example, the lens of the cephalopod eye is proportionately much larger than the lens of the human eye. Moreover, our eye has a blind spot in the middle of the image. The cephalopod eye has no such spot. These and other differences all add up to a cephalopod eye that might perceive the world with a bit more clarity—with the ability to perceive only slight differences in brightness. In the deep sea, where color is less important than light and where animals use light as camouflage, this ability can confer a significant advantage.

Clarity, however, is not to be confused with color. In our environment, the ability to see color is advantageous. Some scientists believe we evolved the ability to see red (most mammals cannot see red) because young red leaves on trees are more nutritious and because red fruits are better to eat.

We humans see color because different wavelengths of light hit three different types of cones in each of our eyes. (Some people, almost always men, are color-blind because they have only two types of cones.) This means that we can see a range of colors that we arrogantly call "visible light." I write "arrogantly" because the term is a little human-centric. Other animals can see much more of the light spectrum. Some, like goldfish and birds, have four types of cones, and a few have even more. Included among this fortunate list are butterflies and, possibly, your basic, everyday, park-loitering, bread-begging pigeon. These animals experience many more colors than we do. Some fish have cones that specialize in paying attention to a deeper, richer red than the red that we see. This means that the fish's ultimate experience of "red" is quite different from ours.

I'm jealous. Imagine the beauty of the Impressionist paintings created by an artist with five types of color cones instead of only our meager three. It's as though we're stuck with watching the world using mid-twentieth-century Technicolor when pigeons, goldfish, and butterflies get to enjoy twenty-first-century digital technology.

On the other hand, most mammals have it worse than we do. They have only two types of cones. While a dog or a cat can see with clarity, it can't see in color. The dog's or cat's world looks luminous, perhaps, but also lacks the energetic "red" that a fish blessed with special types of cones gets to experience. In a dog's world, there are no beautiful shades of red at all.

What humans *do* share with cephalopods (and with many other animals, including dogs), rather than cones, are rods—light-sensitive structures in the back of the eye that are stimulated by shapes and lines rather than by color. So, when we find the color changes in cephalopods so fascinating, it's even more fascinating to keep in mind that the animals making those colors don't directly perceive them. This strikes me as sadly ironic.

But the cephalopods do receive compensation for their loss via their probable ability to see various levels of brightness, or degrees of luminosity. It's worth pausing to think about: Cephalopods make these colors because of the ongoing oceanic arms race and not because they themselves can see them. For survival, it's apparently more important that other animals see the cephalopod colors than that the cephalopods themselves see those colors.

The comparative study of the human and cephalopod eye calls into question "convergent" evolution, the theory that two very different species with very different ancestries might evolve similar solutions to the same problem. The classic example of convergent evolution is the wing of the bat and the wing of the bird. Birds evolved from dinosaurs and bats evolved from the proto-mammals that survived the extinction of the dinosaurs. Yet their wings are similar. Evolutionary theorists used to say that the two wings converged on the same solution. The image is one of two different roads meeting at an intersection.

The human eye and the cephalopod eye were said to be another such example. Those who doubt the theory of evolution have often said that such a thing is impossible and that the similarity of the two eyes is proof of a divine creator. The concept of random mutation resulting in two similar eyes, they suggest, is simply absurd. In that one issue, creationists and Charles Darwin were agreed.

But it turns out that neither eye evolved by accident. The genes to create both eyes, scientists now believe, were probably present from the early days of animal evolution. The cephalopod eye has shown scientists a continuity in evolution that's more organized than we suspected even a few short decades ago.

Indeed, evolution is a much simpler mechanism than anyone guessed. All eyes in the animal kingdom start with the same basic genetic building blocks. These building blocks, certain specific genes, are simply juggled in different ways to make different styles of eyes. Imagine a small set of Legos. From that set, you can build all sorts of things—cars, houses, furniture, railroads. All these seemingly different items evolve out of the same set of building blocks.

This is the surprise: To evolve different styles of eyes, it was not necessary for brand-new genes to appear. All that was necessary was the juggling of genes already present. Eyes throughout the animal kingdom have evolved different styles as a result of complex interactions between this basic eons-old genetic tool kit and the world in which the animal lives. The eye of the nautilus, cousin to the squid, is a simple pinhole eye. It is much less complex, requires much less energy to make and to operate, and is all the animal, protected by a shell, needs to survive. The squid and the octopus, on the other hand, need much better eyes in order to hunt and to hide from predators. You can think of the camera eye as the squid's equivalent of a protective shell: As they lack a shell, the camera eye provides them with a substantial advantage.

All of this might seem a bit far-fetched, but researchers in the past several years have shown that the journey from a simple eye to one like ours is comparatively short—perhaps less than a half million years.

Scientist and author Neil Shubin believes that most animals possess what he calls a "master switch in eye evolution." Because of this, he and others suggest, the concept of "convergent" evolution might be outdated. In a paper in *Nature*, they wrote that the more accurate term might be "parallel" evolution.

· : · : · : · : ·

If it's sad to us that cephalopods can't see the full range of the bewildering beauty they create, their ability to control all these chromatophores with the numerous supporting cells nevertheless requires a great deal of brainpower. The basis of that power is another kind of cell—the neuron—that's apparently also been present for quite a while on the evolutionary timeline. These squid neurons with their tree-trunk-like axons have helped us answer one of the beachgoer's most pressing questions: Why, when a crab bites your toe, does your mouth scream "Ouch!"?

DIAPHANOUS AND DELICATE

The biology of the mind will be to the twenty-first century what the biology of the gene was to the twentieth century.

—ERIC KANDEL

For more than a century, the summertime village of Woods Hole, Massachusetts, has been a world-renowned center of intellectual excitement as well as a fashionable watering hole for the scientific elite. Some of the world's best biologists, including a liberal salting of Nobel Prize winners, have come to sit on the beach and play tennis, to work in the research facilities of the Marine Biological Laboratory, to give and attend lectures, and to exchange ideas. The village's sidewalks overflow with scientists, students, and tourists. There's rarely a place to park your car, even on Albatross Street, and you can count on the Water Street drawbridge, which lets boats leave their Eel Pond moorings for destinations like Martha's Vineyard or Nantucket, being raised and lowered many times throughout the day.

But in the winter, the village can be awfully forlorn. Water Street has a distinctly dowdy look, as though it's down on its luck. Slate-gray skies hang heavy over the silent, institutional buildings. The renowned science library, where you can hold in your own hands scientific journals from the mid-1800s, is almost deserted. If you walk down the main street on the wrong November day, you could easily think that the village is on the skids.

But if you're there on the right November day, there's an intellectual Indian summer. That's when the neurosurgeons arrive, ready to bone up on the latest discoveries in their field. Among the many skills they learn is how to dissect a live axon from a decapitated squid. Or, at least, they *try* to learn that skill.

On the particular day I went to observe, it was chilly and wet. Sheets of rain flooded the streets. Farther north in New England, it was snowing. The first crew of confident neurosurgeons made their way, heads down against the downpour, to the research building where so many Nobel scientists have worked and where squid and other marine life-forms have been studied for nearly a century.

Course teacher Bruce Andersen, a neurosurgeon from Idaho, picked up an eight-inch squid, a common *Loligo pealei*, in one hand. He held a pair of scissors in the other. The animal's chromatophores were showing. It flushed a deep, rich red.

Andersen held on to the squid body. The animal's one head, eight arms, and two tentacles writhed.

"We'll start with the gross dissection," he said.

Then he snipped off the head.

A deep, anguished groan came from the thirty mostly male surgery residents.

"Neurosurgeons are surprisingly squeamish," Andersen told me later.

"And it's all for the good of science," he told them.

"This is all the guts 'n' stuff," he said as he cleaned the body out.

Next, he demonstrated how to lay out the squid's body, find the giant axon that allowed the animal to swim, and gently remove it.

Andersen gave each student a squid and ordered the students to begin their own fine dissection—the removal of the squid axon from the animal's flesh. Properly handled, an axon can continue to function for hours after the animal is dead, even when completely removed from the specimen. The point of the exercise was to remove the axon without harming it.

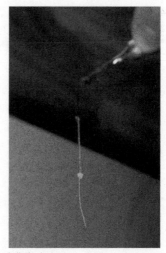

Loligo's giant axon

CHAPTER SEVEN

This turned out to be more difficult than the neurosurgeons expected. *Loligo's* axon is large and easily visible, but it's also diaphanous, like a beautiful bridal veil or a thin sheet of water cascading over rocks. It's as delicate as gossamer and as easily destroyed as the filament spun by a small spider.

Nick the axon cell membrane and you're toast.

All the surgeons tried. All failed.

Their axons died on the operating table.

"You'll all have to go talk to the families now," Andersen instructed. "Luckily, few squid have good lawyers."

. : . : . : . :

It may seem strange that medical doctors practice their neuroscience skills on squid, but it turns out that the squid's neuron with its axon, so diaphanous and delicate, behaves quite like a neuron in our own brains. These nerve cells, or neurons, are "the workhorse of the nervous system," in the words of one particularly articulate neuroscientist, Robert Sapolsky. Without the neuron, we wouldn't function. It allows us to move our muscles, to meditate on the meaning of life, to read books and talk about what we've read. Yet in humans, neurons are ineffably tiny. "Few things in clinical neurosurgery approach the scale and delicacy of dissecting a 300-micron human axon," Andersen said.

We humans have very roughly 100 billion such cells. And as unlovely and nonmammalian as *Dosidicus* and *Architeuthis* and other squid may be, we share this important cell with them. Because of this, scientists suspect that the neuron in one form or another has been around on our planet for quite a while, possibly since the days of *urbilateria*.

For us, neurons are not easy to come by. In general, we get all the neurons we'll ever have soon after birth, although under the right circumstances the human brain may be able to grow a few absolutely spanking new neurons in a few locations in the brain during adulthood. This process is called "neurogenesis," and it remains poorly understood and somewhat controversial. We certainly can't generate new neurons on a large scale, though.

Human neurogenesis pales next to the ability of the cephalopod to continue to create neurons throughout much of its life. In many cephalopod species, if an arm or tentacle is lost, the animal is able to grow a new one. No one understands exactly how this happens, but scientists consider it a fertile avenue for future study.

But there is one overarching and somewhat astonishing truth, a marvelous fact of evolutionary history: A neuron is a neuron is a neuron. The neuron is a near-universal phenomenon, existing throughout much of the animal world. Because life is flexible, there are some differences in neurons among various species, but the basic idea has been around for hundreds of millions of years.

I find this thrilling. Comforting, really. Sharing our neuron— the cell that gives us our individuality and our particular personality—with so many other species makes our planet a little less lonely. The foundation of our ability to think is the same foundation that allows the cuttlefish to change color and shape instantly, or the Humboldt to swim in the ocean or fly through the air at super-high speeds. (Yes, Humboldt and some other squid species can "fly" by shooting out of the water at very high speeds, although they don't flap their fins the way birds flap their wings.) The neuron allows the giant squid to live in the deepest parts of our ocean and the colossal squid to hunt by using its "headlights." It allows birds to navigate our skies. There were neurons in dinosaurs that allowed them to eat, and neurons in the first tiny proto-mammals that allowed them to survive the destruction that killed the dinosaurs and eventually to become—us. As evolution continues and we disappear from the universe, as we certainly will sooner or later, the neuron will probably go on, blossoming in some other intelligent being's brain and, hopefully, creating a life-form that finally figures out how to stop fighting and just enjoy being alive.

· : · : · : · : ·

CHAPTER SEVEN

The neuron is the main cell in the cephalopod's brain, and in my brain, and in your brain. As you read these words, your neurons are hard at work, assembling the black ink on the page into very large concepts, like the universality of life.

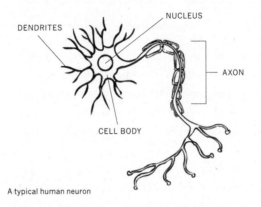

A typical human neuron

The neuron has three basic parts that you need to know about: the cell body, the dendrites, and the axon.

The first is the cell body, a kind of a torso, in a sense. Like your own torso, the cell body contains most of the parts necessary to keep the cell alive. The body of a neuron, an extremely busy place, is like the manufacturing hub of a large city. The nucleus, where most of the neuron's DNA resides, performs a kind of executive function, directing the building of all kinds of molecules the neuron needs in order to thrive and help you accomplish goals like moving muscles, reading a book, and thinking about science.

The second major part of a neuron is the network of dendrites leading into the cell body. Dendrites extend from the cell body like little hairs and are there to absorb information from the world outside the neuron and take it into the cell body for consideration. Dendrites are roughly the equivalent of e-mail in-boxes. Some scientists call dendrites the "antenna systems" of the neuron, because they absorb signals and then send those signals to the cell body. The absorbed information might come from another neuron, or it might come from the world outside. There are special neurons, called "sensory neurons," with unique types of dendrites that allow you to see, to hear, to smell, to touch, and to taste. The number of sensory neurons differs greatly from species to species. A dog does not see the array of colors that we see, but has many, many more neurons and dendrites devoted to smell than we humans do. We are limited to roughly 5 million such smell receptors, while some dogs may have more than 200 million. While the dog misses the glorious world of color that we see, we enjoy only a fraction of the odiferous universe that the dog gets high on when it rides in the back of the car. There are even sometimes immense differences in breeds. German shepherds have twice as many neurons devoted to smell than do dachshunds.

Sometimes the dendrites leading into a nonsensory neuron are so plentiful that, under a microscope, they look very much like a richly branched coral, or perhaps like a piece of finely tatted lace. This is a good thing. In the case of dendrites, complexity is what you want. The more dendrites a neuron has, the better connected that neuron is with other neurons in the brain. Dendrites are constantly growing and changing. The more reading, thinking, information gathering, and just plain experiencing a person enjoys, the richer his or her dendritic connections.

Human infants are born with some, but not a lot, of dendrites. There is, however, plenty of space for dendrites to develop, and if the infancy is normal, that's exactly what happens. The process of growing up is the process of developing more and more dendrites. This is why kids shouldn't spend all their time playing with their electronic toys: They're missing out on building all the other dendritic connections they will need to live a full life as an adult. Experience-deprived kids have many fewer dendrites (and the consequent interconnections with other neurons) than do humans who were fortunate enough to enjoy a highly enriched childhood.

Nurture—experience—intertwines with nature. As important as our DNA is, our lives are not genetically predetermined, because the genes we are born with interact constantly with the world we live in.

The third main part of the neuron is the axon, which performs its main function after the neuron's cell body assembles all the information brought in by the dendrites; if conditions are correct, the information is sent down the axon to another cell. The axon is huge compared to the rest of the neuron. It's often more than 99 percent of the whole cell. Each axon is essential to your body because you don't constantly grow new ones. Moreover, many neurons have only one axon, only one pathway through which they can send information or instructions out of the cell. And that's pretty much it. For life. If you damage that axon, forget about the cell. Eventually the axon will die back all the way to the cell body, and the cell itself will die. Nerve injuries are actually destroyed axons, and researchers have learned that the cause of many neurological diseases is a slowly disintegrating axon.

Neural axons vary greatly in length. The axon may carry information to other neurons located right next door in the brain. In that case, the axon may be very short, perhaps only a little more than a hair's width. Or it may send instructions, like "run away from the crab that just bit you," to muscles all the way down your torso and then connect with other neurons in your spinal cord that pass the message on to your leg and, ultimately, to your toe.

Human axons like these may be several feet in length. A giraffe may have an axon that's as long as 15 feet, while the blue whale, currently the planet's largest animal, may have an axon as long as 60 feet. The blue whale's axon extends from its brain down much of the length of its body. The final result: the flexing of its tail muscles. It takes a blue whale a longer time to send a message to its tail than it takes your brain to send a message to your toe. But the time difference is not one that we'd notice easily, since responses to our environmental surroundings happen, often, much more quickly than we are able to consciously think about them. In other words, we are likely to have already run away from the crab by the time we get around to thinking, *Stupid crab!*

· : · : · : · :

Just as all neural axons are not the same length, neither are they the same diameter. The thickness of an axon is species-dependent. Humans have axons that are too thin to see without a microscope. Instead, what we see are *bundles* of axons, what we commonly call "nerves." The nerve bundle that leads from our eye to our brain, called the optic nerve, has about one million axons. Neurosurgeons rarely operate on individual axons, but instead often work on these bundles when they try to repair nerves. Individual human axons are too small and too delicate to work with under normal circumstances.

But there is one very special group of animals that turns out to have a very thick axon: squid, including little *Loligo pealei*. This tiny animal, which Bruce Andersen held in his hand that rainy November afternoon, is only a few inches in length. Yet it possesses a "giant axon," that, if you look carefully, is visible to the naked human eye. Little *Loligo*'s giant axon is not particularly

A handful of little *Loligos*

CHAPTER SEVEN

long, but it is very thick. It's sometimes said to be "as thick as a pencil lead," but most specimens are not. However, many are about a thousand times thicker than a human axon.

Loligo's axon evolved in order to protect the animal from predators. Squid are among the fastest swimmers in the sea, and the purpose of this very thick axon—also possessed by *Dosidicus*—is to help the squid jet away from danger at lightning speed. Most of us cannot send messages to our arms and legs with anything like the speed of this little squid, although I did that day see Bruce Andersen, after several tries, catch a *Loligo* in a tank with his bare hands. His feat was quite impressive.

For obvious reasons, this thick axon is much easier to study than a human axon. If you are highly skilled—and even neurosurgeons with their super-steady hands need to practice this delicate task—you can remove this axon from the squid and insert tools in it to discover what's going on inside.

This is fortuitous. But what's equally as convenient is that these small animals are abundant. It's not so easy to find a giant squid, but at the right time of year—spring and summer around Woods Hole, as it happens—*Loligo pealei* are as thick as gnats. Traveling only a few miles away from home port, Cape Cod fishermen catch them by the boatload, either to sell to a fish market to be turned into calamari, to use as fish bait, or to send to a research lab for study.

SOLVING FRANKENSTEIN'S MYSTERY

It's the squid they really ought to give the Nobel Prize to. . . .
—ALAN HODGKIN, NOBEL LAUREATE

O n the streets of Woods Hole, there's a fellowship of souls in the peaceful hour of darkness before dawn. "Mornin'" is the preferred greeting at this special hour, along with a polite but reserved acknowledging nod when two bodies pass each other. There's a sense of community. Most of the people up at this hour are finishing their caffeine and heading out on the water.

One promising August morning in 2009, squid lover Joe DeGiorgis, professor of neuroscience at Providence College and a Marine Biological Laboratory researcher who studies the inner workings of the squid axon, was carrying enough coffee and pastry to keep the crew of the fishing boat *Gemma* well buzzed for hours. Joe was looking forward to a fruitful trip. The *Gemma* looks quite like any of the Cape's fleet of commercial fishing trawlers, but it's actually the collecting boat for the Marine Biological Laboratory. For decades, the *Gemma* has gone out most summer mornings with the first raising of the Water Street drawbridge. The mission: collect enough squid, clams, sea urchins, monkfish, and whatever other marine animals are needed to fill the day's research needs of the institution's scientists.

For Joe, this particular morning was a kind of homecoming, since he'd started his career at MBL as a collecting diver whose job was to hunt for animals like the sea urchin and the common surf clam. Along with the cephalopods, these sea animals, like so many other sea species, have contributed greatly to medical research. Both the sea urchin and the surf clam are extremely practical as research specimens, since they are abundant and easy to harvest, and consequently, cheap.

Various sea species have specific qualities that make medical research easier. The eggs of sea urchins, for example are large and develop quickly. They're also transparent, so that, unlike in a hen's egg, for example, scientists can watch the process of the animal growing and developing inside the egg.

Victorian scientists used sea urchin eggs to learn about human fetal development. In 1899, MBL summer scientist Jacques Loeb made an important discovery: He could get unfertilized sea urchin eggs to divide and form new animals by putting the eggs in certain kinds of liquids. The ease of initiating the development process meant that researchers could—and did—make astounding progress in understanding this very early stage of the development of a living organism.

Surf clams, the diced-up bits of flesh you're probably eating when your dip your spoon into a bowl of clam chowder, have also contributed to the progress of human medicine. Surf clams are mollusks, like squid, but much less complicated. The surf clam's life consists mostly of hiding in its shell, buried deep in the sand, sucking water in and filtering out whatever nourishment is around.

While Joe was a diver, scientists were using surf clams and sea urchins to do basic research that ultimately ended up improving cancer treatments.

Almost all cells in the human body divide into two, then regrow. This is why skin heals, fingernails grow, and hair gets long. Growing new cells is an essential process: Out with the old, bring in the new. But the body usually controls that process very carefully. Under normal circumstances, the cell divides only after a certain time period.

Promiscuous activity resulting in lots of cell division is unwelcome. Cells are not supposed to keep dividing. Various types of cells in your body routinely divide at different rates, but most of the new cells are supposed to undergo a kind of rest period, before they wear out and are cast off. A cancer occurs when the cells continue to divide and divide, never taking time to relax and just smell the roses.

To better understand why some cells become cancerous, researchers need to better understand the basic biology of cells.

What makes normal cells divide in the first place? Without understanding the normal process, it's harder to control the abnormal process. Using some of the species of animals collected by Joe, researchers learned that two different complex molecules in a cell—called "cyclin" and "ubiquitin"—control the basic divide-then-rest routine.

Cyclin and ubiquitin are a yin-and-yang pair. They work as a duo, balancing each other out in a wonderful example of teamwork. Cyclin builds up over the life of a cell, then, when the correct time comes, the ubiquitin attacks the cyclin. It breaks up the complicated molecule so that when the cell divides, the new cells don't have as much. Then the cycle of buildup and breakdown begins anew in the just-divided cells.

Nature is often organized in this pleasantly logical way. The compound cyclin was discovered by MBL summer scientist Tim Hunt of Great Britain in 1982, while he was studying sea urchins. Hunt learned that the amount of cyclin in a cell increases gradually over the life of that cell. Finally, it reaches a peak. The peak is the signal for action. The cell divides. For finding this clue in the mystery of why cells divide, which provided a completely new strategy for treating cancer, Hunt and several colleagues won a Nobel Prize in 2001. Later, Joan Ruderman, one of the students who worked with Hunt in 1982, found that some breast cancer cells do indeed have too much cyclin, which could possibly initiate too much cell division.

The discovery of ubiquitin came around the same time. Ubiquitin is so named because it is ubiquitous—present, like cyclin, in nearly all animal cells, even in tiny algae. It is ubiquitin that destroys cyclin so that the new cells don't have too much. What goes up must come down. If the amount of cyclin were high in the new cells, the cells would keep dividing. Ubiquitin ensures that this doesn't occur by breaking apart the cyclin, so that the healthy, well-measured dividing of cell growth can begin again. The discovery of ubiquitin earned MBL summer researcher Avram Hershko (who, as a six-year-old, narrowly escaped an Auschwitz gas chamber) and several other colleagues the Nobel Prize in 2004.

One of the most marvelous things about cyclin and ubiquitin is that these molecules are present in almost all living cells—in plants, in yeasts, in most animals, and in humans. Across all these species, the compounds are similar enough that scientists believe they must have been present in very early life-forms. In scientific jargon, the genetic recipes for these molecules have been "conserved" throughout much of evolution.

· : · : · : ·

While Joe was diving for research purposes, he began to learn about the natural behavior of squid. He learned that *Loligo pealei* has elaborate courtship behaviors, and that when a male mates with a female, he sticks around afterward, trying to keep the other males away.

"It's like a bar scene, when a guy has one eye on his girlfriend and another on every other guy in the room," Joe explained.

When the first female of a shoal of squid lays her eggs, the second comes along and lays her eggs in the same place. The second female attaches her eggs to the first batch of eggs, and so on and so forth, until all the females have left their gifts to the sea. "It finally looks like a huge anemone," Joe said. "There are hundreds of fingers, all containing eggs, physically linked together. It's an event. They reproduce as an event." From then on, the development of the baby squid is synchronous. They develop together, watching with their eyes all the others in their group. They hatch together. They school together, swimming as a group throughout their very short life span (less than a year). Thus, one shoal of hundreds of squid may go through a complete life cycle together, as though, in some ways, they are one living being.

Joe was also fascinated by the braininess of the little animals. "Besides the fact that they're very beautiful, they're very intelligent," he said. "The point is—they're thinking. Does a mouse think on their level? Probably not. Does a dog? Depends on the dog."

For a while Vineyard Sound *Loligo* were in great demand in laboratories around the world. Joe started a business called

Calamari Inc. Laboratories put in their orders for a variety of squid parts—eyes and axons and fin nerves and brains—and Joe would dissect the squid and send the scientists what they needed. A scientist from the National Institutes of Health called him one day and asked for 2,000 squid eyes. The scientist was studying how eyes actually see. Joe sat down and removed squid eyes, one every five minutes at $5.00 an eye, froze them, and sent them to Washington.

At $5.00 an eye and twelve eyes an hour, Joe was making pretty good money for a kid. But he eventually realized he was bored sitting on the sidelines. He was interested in the neuroscience itself. He went on to earn a doctorate in neurobiology. These days, he heads his own MBL lab, and is working on squid science that he hopes will help lead toward a cure for Alzheimer's disease.

But he's always happy for a chance to ride on the *Gemma*. As we crossed the Sound that August morning, basking in the early morning sun, we headed for a prime fishing spot near the island of Martha's Vineyard. I asked DeGiorgis about his most unusual dive in these waters when he was still working for the collections department. He said it was the sudden appearance one day of hundreds and hundreds of salps, strange jellyfish-like organisms that make long chains and float through the water, eating and growing along the way. On that particular dive, he had to part the strands as he moved through the water. "It was like walking through a beaded curtain," he said. "It was a virtual sea of salps. Then, the next day, they were all gone. Vanished. It was a place where I've dived more than anywhere else in the world. I've never seen them again. Bizarre."

Joe DeGiorgis dissecting a squid

The squid fishing wasn't great that warm late-summer

day, but after a series of runs with the trawling nets, the crew had brought up enough *Loligo pealei* to call it a day. Back at the dock, scientists and their lab assistants dropped by to pick up their orders for another day of research.

After the trip, Joe dissected an axon to show how it's done.

"The neuron is shaped like a tree," he explained later. "It has 'branches'—the dendrites; a 'trunk'—the axon; and 'roots'—the terminal end of the axon. The axon, the trunk of the tree, is what I'm interested in."

He tied off both ends of the axon using thread, black on one end and white on the other, like you might tie the end of a balloon. Then he gently lifted the axon out of the squid's body and placed it on a petri dish. He peeled away the unnecessary tissue clinging stubbornly to the outside of the axon. It was kind of like peeling a banana. That left the naked axon, containing only the goo—the "axoplasm"—that filled up the axon's insides.

It took him about five minutes. In Joe's experienced hands, the task looked easy.

To prove the axon was still doing its job, he put an electrode inside it.

The clicking sound, the buzz of electricity, was clear as a bell.

$$\cdot \; : \; \cdot \; : \; \cdot \; : \; \cdot$$

As nerves, bundles of axons produce the river of power that runs through your body. Electricity, the same physical force that turns on your electric lights or makes your computer work, is the force that enables you to think and dream and play baseball and drive a car. That's why many medical textbooks equate the nervous system to a system of electrical wiring. "Life exists because of a delicate dance of electrons," wrote author Joseph MacInnis. Without *Loligo pealei*, we might not have several basic facts that led us to this understanding.

For thousands of years, we've known that some sea animals produce electric shocks. Torpedo rays, capable of stunning their victims with as much as 220 volts of electricity, found their way into Plato's *Dialogues*. Early Roman physicians used these animal-generated electric shocks to treat human ailments like headaches and gout, presumably with some success. But while ancient cultures understood that some animals were capable of emitting shocking levels of electricity, they did not understand that *all* muscles—including our own—contain and discharge electricity.

The fact that our muscles work because of electricity was discovered around the time of the American Revolution. Other scientists had played around with the phenomenon of electrical interaction with animal muscle, but it was Italian scientist Luigi Galvani who did the first series of solid experiments in the field, in the 1780s. During a storm, Galvani saw that a severed frog's leg, hung outside on a copper hook on an iron balcony, twitched when lightning appeared in the sky. He also found that he could make the severed leg twitch with static electricity, as well as when he sent electric current from a Leyden jar, a kind of very primitive battery, into the dead leg. He thought about this a great deal and finally suggested that the muscles themselves created their own unique kind of electricity, which he called "animal electricity."

While Galvani was studying "animal electricity," other scientists were independently studying electricity more generally. Ben Franklin, of course, established that lightning bolts were bolts of electricity, and also coined many of today's basic electrical terms, like "positive," "negative," and "current." The impish Franklin liked to show off for his dinner guests by killing turkeys with bolts of static electricity.

Franklin and other researchers imagined that electricity was rather like water, only invisible. Yet while Franklin and others were able to work out a bit about *what* electricity did, they did not understand *why* electrical phenomena occurred. The discovery of what was actually flowing—energy from the bouncing around of negatively charged electrons—would not occur for another century.

At first glance, the discoveries of Galvani and Franklin and many other scientists seemed contradictory. Too much electricity could, obviously, kill. On the other hand, a jolt of electricity

seemed, from Galvani's experiments with frogs, to give life.
How could this be?

Scientists proposed the existence of two different kinds of
electricity and suggested that the electricity in a frog's muscle
differed in some basic way from the electricity Franklin discovered.
The electricity that moved muscles became "animal electricity."
Franklin's lightning-bolt electricity became "natural electricity."

This may seem silly to us today, but then many scientists
thought it reasonable. After all, how could bolts of static
electricity kill a turkey but also, apparently, give life to a dead
frog's leg? The whole thing seemed very odd.

The public was both confused and fascinated. An "electrical
frenzy" swept Europe, writes neuroscientist and historian Stanley
Finger. For much of the nineteenth century, people imagined
that electricity could do all kinds of things. Percy Shelley, the
great English poet, tried to cure his sister's skin disease by using
electrical shocks and, explained Finger, "managed to electrocute
the family cat in the process."

Smatterings of what the scientists had learned gradually
entered the popular culture. The verb "to galvanize" entered the
vernacular, and some people claimed to be able to use electricity
to encourage people to become more active. (And indeed, it does
turn out that when you administer an electric shock to someone,
you do "galvanize" them into action.) Other people began to
wonder if the electrical force wasn't what created the human
"spirit," which seemed to disappear when a person died. Could
you use electricity to bring back the dead?

As often happens, this scientific breakthrough caused a
popular uproar. To many people, it seemed as though scientists
were eating apples from the Garden of Eden, usurping knowledge
and abilities that should belong only to God. And thus was born
Frankenstein; or, The Modern Prometheus, perhaps the first-ever
novel to be written in the mad scientist genre. Mary Shelley, a
bored English teenager, was hanging around the resort of Lake
Geneva with friends, including the poet Lord Byron and her lover,
Percy Shelley, in the extremely cold and very rainy summer of
1816. Because of the weather, Mary and her companions were

stuck inside, forced to huddle around a warm fireplace, swaddled in layers of clothing.

On one of those chilly evenings, Mary listened to Percy and Lord Byron explore the essence of the word "galvanize." Was it really possible, they wondered, to assemble body parts and create a living being? Would it ever be possible to discover "the nature of the principle of life"? In Mary's fertile imagination then appeared the fictional character Frankenstein, a scientist who did just that. In her novel, after building the technology and assembling the body parts, the professor brought to life a humanlike monster. "It was on a dreary night in November," Mary wrote in chapter five, "I collected the instruments of life around me, that I might infuse a spark of being into the lifeless thing that lay at my feet." Powered by electrical shocks, the monster opened its dull yellow eyes, breathed, and began to move its muscles.

It wasn't until the end of the nineteenth century that scientists finally resolved the confusion. The physicist J. J. Thompson proved the existence of electrons, particles even smaller than atoms. Scientists then understood that what was flowing through nerves and axons were really electrical charges created by the activity of these electrons.

If the discovery of electrons helped physicists understand a bit more about electricity, it also helped neuroscientists get back on track regarding the work of the brain's neurons by putting to rest the idea that electricity could bring the dead back to life.

· : · : · : ·

Enter *Loligo pealei*. In the middle of the 1800s, scientists discovered that an electrical impulse travels along an axon at a speed of about 90 feet per second, much more slowly than it would travel through a metal wire. By that time, scientists were able to understand the concept of electrical flow through a wire, and imagined that the body was filled with continuous "wiring" that seemed to have something to do with these long strands of fibers that ran down the spinal cord. It wasn't until the 1880s that a Spanish scientist was able to draw the neuron and explain what the various parts of the neuron, including the axon, actually did. By finally clarifying that the neuron with its axon is the basic unit of the brain and that the body does not have a continuous system of wires running through it, Santiago Ramón y Cajal became one of neuroscience's most famous researchers.

Progress continued, although slowly by the standards of modern science. Once science accepted that the neuron had a beginning (the dendrites), a middle (the cell body), and an end (the tip of the axon), researchers were able to learn that all electrical impulses traveling along an axon have exactly the same strength. That is to say, there are not some very powerful electrical pulses and some very weak electrical pulses flowing along an axon. This seemingly uninteresting fact had considerable implications: It meant that the electrical message being sent down the axon was charmingly simple. It was binary, like the telegraph. Either dots or dashes, on or off. Or in terms of computer language, either zeros or ones. The pulse either traveled down the axon—or it didn't.

When scientists attached some simple technology to human nerve fibers, they could hear a characteristic "buzz" when electrical messages traveled down those nerves. But they still couldn't understand the details of what was happening. One of their biggest handicaps was that they were not able to look *inside* a human axon as it was firing. It was simply too small and too delicate for the technology that existed in the decades after Cajal's momentous discovery. For decades, scientists were stumped. It seemed as though it just wasn't going to be possible to study the interior workings of a neuron. How the cell did what it did was apparently going to remain a mystery.

Then British biologist and cephalopod fanatic John Zachary Young came across the Atlantic in the summer of 1936 to enjoy Woods Hole. Young had been interested in cephalopod anatomy from the beginning of his career. That summer at the Marine Biological Laboratory, he worked with *Loligo*.

Young began studying a long, delicate strand of squid tissue believed by most scientists to be a blood vessel. Young stimulated one end of this bit of tissue and heard the characteristic static that showed that electrical impulses were traveling down the pathway. He determined that the tissue was not a blood vessel at all. It was actually a very large axon.

Back in Britain, he continued his research for a while, but not for long enough. Call it a quirk of fate. He decided not to pursue this line of research to its ultimate end—figuring out how the cell managed to create the electrical impulse and send it from one end of the axon to the other. Thus this great scientist gave up an opportunity to earn a Nobel Prize.

He instead handed the research on to Alan Hodgkin and Andrew Huxley. Hodgkin had worked on squid with Young, and he asked Huxley, his onetime student, to work with him to continue to solve the mystery. The two formed the exceptionally powerful bond of two scientists who work well together. Their first job was to develop an appropriate approach. Removing the axon from the squid, they placed it in seawater. They took a finely honed electrical probe and placed it inside the axon and placed another probe outside. They already knew from earlier experiments by other scientists that the inside of the axon, when at rest, had a comparatively negative charge and that the outside of the axon had a comparatively positive charge. Scientific consensus theorized that the charge inside the axon would move from negative to neutral.

Instead, Hodgkin and Huxley discovered, much to their amazement, that there was a big jump inside the axon, much bigger than anyone had anticipated. In fact, the inside of the axon changed to strongly positive. By comparison, the outside became negative. Then, when the axon returned to its resting state, the charges returned to their original charges.

"We have recently succeeded in inserting micro-electrodes into the giant axons of squids," the scientists wrote triumphantly in a short note to other scientists, published in *Nature,* a prestigious British science journal. The pair mentioned in the note that although they had succeeded in one respect, there were many questions they still wanted to try to answer.

Ultimately, they were able to watch the flow of electricity

down the axon as though they were watching the flow of a twig down a stream. They realized that one type of charged molecule moved from outside the cell axon to inside the cell axon, while the other type moved from inside to outside.

Then the pair found something even more intriguing. While many of the molecules involved in keeping a cell alive are rather large and complicated, those involved in keeping the electricity flowing were comparatively simple and very common: sodium (like the sodium in table salt or seawater) and potassium (found in foods like tomatoes and bananas). Both the sodium and the potassium lacked one electron, so they became positively charged "ions." When the axon is at rest, the scientists found, there are a lot more positive ions outside the cell than inside the cell. When electricity flows down the axon, some of these ions outside the cell are in fact moving into the cell. When the electrical impulse passes, the ions move back outside. The inside of the axon returns to its original, comparatively negative state.

By studying the giant axon of little *Loligo*, Hodgkin and Huxley had made this profound discovery: Our ability to think is based on this marvelously simple process—the movement of electrical charges, ions, into and out of the axon.

But like many scientific discoveries, this one raised questions. Why were potassium and sodium moving into and out of the cells at only the appropriate times? Why didn't they move back and forth randomly?

In turned out that there were gates, or channels, in the axon that opened and closed at only the appropriate times. In science, the more questions you answer, the more questions materialize. This is part of the fun. Once Hodgkin and Huxley revealed the basics of how electricity flows down an axon cell wall, other scientists wanted answers to questions about exactly how these gates or channels operated.

Scientist Clay Armstrong, one of Huxley's students, tackled the question. Armstrong sometimes used the "giant" axons of the little squid found near Woods Hole, but he also traveled to South America, where fishermen provided him with Humboldt squid.

Armstrong discovered that individual ions like potassium have their own specific gates that open and close only for them. These gates control passages through the cell wall that have come to be called "ion channels" and that are voltage-sensitive. In other words, the flow of electricity down the axon involves the opening and closing of these various channels.

This complicated-seeming idea is quite simple: Imagine a field filled with horses and cows. The horses can only enter and leave by one gate; the cows only enter and leave by another gate. "We are what we are because of ion channels," Armstrong explained to one interviewer. To another, he explained that "every perception is encoded in electrical form. All of our thoughts, all of our emotions, involve the action of millions of ion channels. Billions."

Since then, scientists have learned that there are many different kinds of channels leading into and out of an axon, and, indeed, into and out of all kinds of cells in the body. The reason some tranquilizers work is that they block the flow of ions through these channels, and thus the flow of electricity down the neuron's axon. The axons or nerves become quiet.

Armstrong's squid-based discovery has had immense consequences for human medicine. A whole new class of medications—channel blockers—has saved countless human lives. A common channel blocker called a calcium channel blocker is routinely prescribed to lower blood pressure and prevent heart attacks. Other medications help to control some forms of diabetes by influencing the opening and closing of potassium channels. Some forms of epilepsy seem to be the result of the malfunctioning of the ion channels in neurons; researchers hope to eventually find medications to improve life for epileptics by controlling the channel malfunctions.

In fact, Armstrong's work on squid has led to a whole new field of medical research—the study of channelopathies, or the study of the malfunction of channels in the axon. It's not hard to see why Hodgkin and Huxley's discovery using *Loligo*'s giant axon has been called one of the most important breakthroughs in the twentieth century. The pair received the Nobel Prize in

1963. Many people expected Clay Armstrong also to win a Nobel, but sadly, he was never so honored.

It seems incredible to me that nature has worked out such a system, so consistent across species, at once brilliantly simple—based on the tiny ions of sodium and potassium, and on simple binary code—and yet so complex, in that it controls so many different processes in our bodies. And yet this is why we can think, birds can fly, and cephalopods can change their colors in only milliseconds.

SERENDIPITOUS SQUID

Chance favors only the prepared mind.

—LOUIS PASTEUR

Around the time Ben Franklin was killing wild turkeys with electricity in the colonies, Horace Walpole, an English public intellectual and the Fourth Earl of Orford, was contemplating the phenomenon of accidentally finding out about things you weren't necessarily trying to understand. Walpole realized that these accidental achievements were more common than you might think—common enough, in fact, to deserve their own unique term.

Thus did Walpole coin the word "serendipity." There is more serendipitous science than you might at first suspect. Until the early 1950s, scientists mistakenly believed that humans had forty-eight chromosomes in their cell nucleus. Then a solution of chemicals accidentally spilled on a dish of human cells. Soaked by the unique chemical solution, the chromosomes swelled and were each clearly, individually visible for the first time. It turned out that humans have forty-six chromosomes, two less than was thought. Moreover, by making individual chromosomes easily visible, the scientist, T. C. Hsu, paved the way for medical research that would eventually save the lives of countless people suffering from chromosome-based diseases.

The most often cited example of serendipity involves the Scottish biologist Alexander Fleming, who won a Nobel Prize for discovering the curative abilities of penicillin in 1928. The commonly told story is that Fleming discovered the organism from which penicillin is made. He didn't. Other scientists had seen the fungus before. But it was Fleming who realized the importance of what he was seeing and who did something about it. It was Fleming who discovered that *Penicillium* fungi could be used to cure an infection of *Staphylococcus*, a bacterium often deadly to humans. Fleming is therefore considered the founder of the field of antibiotics.

The story goes this way: Returning to his lab after a brief trip, Fleming saw that he had forgotten to put away a petri dish containing the *Staph* he had been studying. When he looked at the dish, he found that some of the bacteria had been killed. He also saw that some other life-form was growing there instead. It turned out to be a fungus. He began working with this material and, after much persistence, created the world's first antibiotic medicine—penicillin.

Which goes to show: Serendipitous discovery isn't entirely accidental. You have to be in tune enough with what you're looking at to know that you're seeing something important.

· : · : · : · : ·

"My brain feels like Jell-O" is sometimes used, tongue in cheek, to describe a feeling of mental exhaustion, but in fact scientists do use the word Jell-O to describe the texture of the goo inside your axons. "You can pick up clumps of it with forceps," Joe DeGiorgis told me, "and you can squeeze it out of the axon the way you squeeze toothpaste out of a toothpaste tube."

When Joe was in high school, the axoplasm in a neuron was described as a "soup," he said, "but it's not like that really. It's thicker. You can pick it all up, and it stays together." The material is a fluid, but a very sticky fluid, perhaps just a bit thicker than Jell-O. It's so difficult to describe that many scientists have adopted a highly technical description—"goo." The term appears not uncommonly in the scientific literature.

While some scientists were studying the flow of electricity along the axon, others were looking at what went on in the goo. How did the "Jell-O" function? What, exactly, was this plasma? What kinds of molecules were in there? If most of the manufacturing and maintenance work occurred in the cell body, under the direction of the executive DNA in the nucleus, how did the packages of information get "mailed"? How did food—

that is, energy—get from one place in the neuron to another? One purpose of the axon is to send an electrical pulse from one point to another, but many other support functions also need to happen in your neurons in order for you to be able to contemplate the words printed on this page.

· ∶ · ∶ · ∶ ∶ ·

Scientists have known for quite a while that the axon was filled with a gelatinous substance, and speculated quite reasonably that the substance must have some important job in helping us think. By the end of the nineteenth century, they were able to use simple materials to stain the insides of the neurons and look at a few of the structures there. They could, for example, see the DNA in the cell nucleus, although they had no idea how it worked.

They could also see, with the proper technology, the strange, tiny, sausagelike mitochondria, where energy is made ready for the cell to use. When you eat a jelly bean for breakfast, your body does all kinds of things with that food, but ultimately some of that energy reaches your neurons. In the neural cell body, a portion of that energy goes into the mitochondria, which are little power plants. In these power plants, the energy of food is changed into ATP, which is the form of energy that your cells need to keep functioning.

Some cells in your body have only a few mitochondria, but neurons contain hundreds. This is why kids need to eat breakfast before they go to school: Without food, the mitochondria remain unemployed and kids don't have the mental energy—ATP—that helps them think.

Although scientists had known for quite a while that these structures existed inside the neuron, they had little under-standing as to what exactly they were or why they were there. The only way to see them was to kill the cell, stain it, then study it under a microscope. You could see the mitochondria along with other structures, but you could not watch them work. As microscopes became more and more powerful and as staining techniques improved, scientists could with ever greater clarity look at the structures inside the neuron. But they could never see those structures *moving*—never see the living manufacturing plants, or watch their packaged exports travel through the axoplasm.

In the late 1940s, medical research, spurred by the nerve injuries of World War II, concentrated on improving the understanding of how axons worked. A pair of researchers performed an impressively simple experiment. They took some silk and tied it around a bundle of axons, a nerve. After creating this "dam," they saw that the part of the axons closest to the cell bodies gradually "ballooned." Looking at this bulge on one side of the silk tie but not on the other, they realized that something *inside* the axon was flowing, just as the electrical pulse flowed down the cell wall. So it turned out that there were at least two flow systems in the axon—the flow of electricity and a flow of axoplasm, with its various smaller structures, within the axon itself. The flow of axoplasm inside the axon, however, was much, much slower than the electrical pulse. Researchers estimated that the axoplasmic flow might be only a millimeter or so a day. In comparison, the electrical pulse zips along the cell wall.

By the second half of the twentieth century, scientists could use radioactivity to follow some of the movements. But the details were still mysterious. Exactly how did these packages— tiny organelles, including the sausagelike mitochondria—move? Did they just drift along? Was there some kind of system? It clearly wasn't just random chance that got essential proteins and energy from one part of the cell to another. But how organized could something like that be?

In the 1970s, the field of cell biology was somewhat stymied, in part because the tools available to researchers were not up to the task. Light microscopy—the traditional kind of microscope that we used in high school that relies on visible light—had gone about as far as it could go, at least when it came to resolution limits. At a certain point, when scientists tried to look at smaller components inside a cell, the image would become confused. It was somewhat as though you were looking at those old-fashioned stereoscopes when the tool wasn't the right distance from your eyes.

I called up Scott Brady, an expert on the secret life of the axon. Now a senior scientist and lab head at the University of Illinois, Brady in the early 1980s was a young and ambitious researcher spending summers at Woods Hole.

Progress on understanding the axon had come to a standstill, he told me, because the light microscopes of the day were inadequate. "You started not being able to tell whether it was one versus two objects you were looking at. We were basically stuck, because the kinds of questions we wanted to explore were below that size limitation."

Some researchers suggested adopting the newest video technology, but many of the older scientists remained skeptical. "The reigning paradigm was that TV would never be something you would want to use," cell biologist Nina Strömgren Allen told me. Strömgren Allen was part of a group of researchers who would overturn that paradigm. Her father, a Danish astrophysicist and formerly a student of Neils Bohr, told her about the new high-powered, high-resolution telescopes that had recently been developed for astronomy. Could some of that technology be transferred to the world of microscopy? Strömgren Allen and her husband, Robert Day Allen, began working on a new idea—Video-Enhanced Microscopy—that they hoped would be able to show the inner life of a living neuron. By 1979, they had made some progress, and by 1981 they had published a paper on their breakthrough.

· : · : · : · :

Then one of those key moments of serendipitous science occurred. The couple was teaching a course in Woods Hole about how to use microscopes. They placed a common *Loligo* squid axon under a light microscope, to which was connected a video camera and a television screen. When they switched on the technology, the image didn't seem quite clear.

Allen turned a knob, hoping to bring the image into focus. Twisting the knob controlled how much light was let in and, therefore, how much you could see. Allen was explaining this principle to his students, and he wanted them to see the problem.

Continuing to look through his microscope, he twisted the knob. "See," he said. "You can't see anything anymore."

Hands waved in the audience of students.

"The image isn't washed out. We're looking right at it," the students said. "In fact," they continued, "we can see it *better* than before."

"Of course it goes away," Allen said. "I can't see anything."

Then he stepped away from the eyepiece of his microscope and looked at the image on the television screen.

Suddenly, right in front of everyone, appeared the movement of tiny little components *inside* the squid axon. A whole new world was revealed. Word of the amazing sight spread quickly on the streets of Woods Hole. Scientists passing each other on the sidewalks buzzed with excitement.

"Suddenly, the veil was lifted," Strömgren Allen remembered when we spoke on the phone. "We knew they were there, but we couldn't see them before this."

When Allen followed up on what had happened, he found out that the new microscope that had been sent to him had been tuned incorrectly by the manufacturer. Serendipitously, it was this incorrect tuning that revealed for the first time this whole new avenue of exciting scientific research.

· ː · ː · · ː ·

The accident created a revolution in medicine, just as the creation of penicillin had decades earlier. Scott Brady was there the day it happened. He said: "It provided us with a means to visualize these very tiny objects in living preparations. You could resolve things that were smaller with electron microscopy, but you also had to impregnate what you wanted to see with metals, so that you weren't then seeing much in the way of direct objects, but depositions of metal stains. All of that was dead, and if you're interested in movement, it's not going to do you much good. It was really quite remarkable. No one knew there was so much movement. And no one had realized that all this movement was almost continuous."

Suddenly you could look at a lot more than the electrical pulse flowing along the squid axon. You could see that *inside* little *Loligo*'s axon was a beehive of activity. Or, to use the analogy provided in one scientific paper, it was "as engrossing as the ant farms of our childhood."

Scientists had never imagined that the world inside the axon was so dynamic. And the first thing they noticed—what was stunningly obvious—was that this activity was much more than just a lackadaisical "drift" of organelles. There seemed to be roadways and pathways, lots of stop-and-go traffic, and much more organization that anyone had previously imagined. There was a whole universe in there. It was as though you were looking at a model train set, the very elaborate kind you see in department stores at Christmas. There were engines going in all sorts of directions, lots of different tracks, and loaded-up flatbed cars and stop-and-go points where things were loaded and unloaded. It was all highly orchestrated so that (usually) none of the moving parts, the molecules, collided with each other. Some of the engines seemed to be zipping along, much faster than the one millimeter a day that had been estimated. Others crept along at a pace that would have made a snail look high-powered.

There were even, occasionally, accidents. Sometimes scientists could see collisions. Every once in a while, the molecules pulling their loads would inexplicably (or so it seemed) jump the tracks.

The movements the scientists saw that day were breathtakingly sophisticated. "It was a jaw-dropping experience for those guys, and started a whole new flurry of activity. It's a very famous story," Joe DeGiorgis explained. Worlds within worlds, right there, in each and every neuron. Many of these organelles were being tugged along trackways, out of the cell body and into the axon, traveling some-times all the way to the axon tip. In the 60-foot blue whale axon, this is quite a trip.

· : · : · : · :

What soon became obvious was that various molecules had specific jobs. Using the new technology, a number of younger scientists began studying the intricacies of the activity. Intriguingly, they learned that they could squeeze the axoplasm out of the squid axon, and that, under proper conditions, it would continue to do its job of shuttling the tiny packages around for quite a while. This made their task much easier, because they could look directly at the tracks and cars, rather than having to look at them through the semitranslucent cell wall.

Almost immediately the scientists began looking at the chemistry of the miniature railroad system. Scott Brady was in on the action right from the beginning. The senior scientist he was working with in Woods Hole told him about the news as soon as it happened. For a young scientist, to be present at a revolution is like receiving manna from heaven. This moment of video-enhanced clarity provided Brady, in an instant, with his life's work, with a whole new wide-open field of research where no other scientist had yet staked out territory.

It was like being the first prospector to find gold at Sutter's Mill.

"There was so much excitement," Brady remembered. "Then we started asking questions: How can we make use of this?"

The race was on among the young scientists in Woods Hole that summer to be the first to find out what the various molecules inside the squid axon were up to. Thrilled with the new technology, Brady set about trying to find how the larger molecules were pulled along the tracks in the axon.

Both Brady and a competing young researcher, Ron Vale, found a kind of "motor," eventually called "kinesin," that was responsible for moving packages up and down the axon. It turned out that kinesin was a kind of "slave" molecule that "walked" along one of the tracks, putting one foot in front of the other, pulling its loads behind it.

Eventually, after decades of research, researchers, including Joe DeGiorgis, had found many different kinds of hardworking kinesins in squid and other animals, including humans.

There are multiple motors in the same neuron. Humans may have nearly fifty different kinds.

I asked Joe why we need so many.

"We're trying to figure that out," he answered. "We don't know yet what all these motors do. We know it's a trafficking issue. We want to know that if there's a problem with some of this transport, does that lead to neurological disease?"

$\cdot\ \vdots\ \cdot\ \vdots\ \cdot\ \vdots\ \cdot$

So, inside the axon is a city that never sleeps.

Thanks to the squid, we understand this. But how does an axon die? What happens to brain cells when we develop diseases like Alzheimer's or Parkinson's? Brady and a number of colleagues across the country and around the world have devoted much of their research during the first decade of the twenty-first century to answering that question. And once again, they've used the squid axon in some of their research.

"Neurons have some very special challenges," Brady explained. "You have to remember that neurons are, many of them, extremely long. When you start stretching a cell over a meter [a meter is a bit more than three feet] or more, we're talking about large as well as long. Especially when all the proteins needed all along the axon have to be packaged and transported to where they're going."

Recently, scientists have learned that the trackways inside the axon run in both directions. Packages put together in the cell body must sometimes travel all the way to the end of the axon. And materials at the far end of the axon sometimes must travel all the way back to the cell body. This two-way transport is mandatory. It's also sometimes mandatory for the packages to be dropped off at points in between both ends. "These things are essential for the survival of the cell. It turns out that you have to have particular proteins at particular places all along the neuron," Brady said.

So, Brady and others wondered, what makes the proteins inside the neurons start and stop? How do the kinesins "know" when to dump their loads and relax? How do they know when to keep on truckin' just a little bit farther? One of the triggers, or switches, that flips the kinesins on and off turned out to be another molecule, called a "kinase." It's the kinase's job to give kinesins their marching orders. Kinases are like the switchmen positioned along railroad tracks, managing traffic.

"What happens when the kinase doesn't do its job?" I asked.

"You get a perfect storm," Brady answered. "The axon starts dying back. And if the axon dies back far enough, then the whole neuron dies."

In 2009, Brady and others published papers that tied at least part of the problem in patients suffering from various types of neurological diseases to the malfunction of the kinases in the neuron that give the kinesins their orders.

It reminded me of the story Yale's Vincent Pieribone had told me about the different characters—"amazing little guys"—inside the axon. Using squid and other animals, scientists keep finding more and more such characters, all with their own individual, highly specialized jobs to do.

By studying kinases, the switchmen, Brady and his lab team believe they have discovered an important part of what happens in the human axon when Huntington's disease debilitates a human body. Researchers have long known that the disease has a genetic, inherited basis. Because of this genetic problem, a long chain reaction or a cascade of errors occurs. The kinase does not do its job properly. Which means the kinesins do not keep on truckin'. The packages that need to be carried back and forth between the tip of the axon and the cell body do not get where they need to go in the correct quantities or in the correct time frames. Eventually, the very existence of the neuron itself becomes an issue.

Brady believes that there are a number of common neurological malfunctions that have their roots in the malfunction of the axon's shuttle system, and that Alzheimer's and Parkinson's disease may be among them. He and his lab have even created a name for this group of illnesses: dysferopathy.

· : · : · : · :

It took several decades and many, many scientists to decipher this puzzle, and the success of the endeavor stretches all the way back to J. Z. Young and the discovery of the giant axon in little *Loligo pealei.* It's been nearly a century of scientists standing upon the shoulders of the generation that preceded them.

I asked Brady about that history, and about how odd it seemed to me that so much of our own brains have been revealed by studying an animal that's so incredibly different from us.

"The squid was designed by Mother Nature for neuroscientists by making everything so big that it allows us to see things and have access to things that we just really can't get to in an intact mammalian system," he answered. "These kinds of experiments can only be done with squid. We share a surprisingly large number of features with the squid. The kinesin motors, for example. There are big chunks of our kinesin motors that are remarkably similar to those of the squid. We both have the same basic mechanisms. The choices [about how the neuron would develop] were made before we split [on the evolutionary tree], perhaps 700 million years ago. And we both took advantage of those choices to re-create the remarkable signaling that is the nervous system. So far, everything that we've identified in the squid, we've been able to confirm in the mammalian system."

I asked him the question I asked everyone: If squid have such complex brains, are they smart?

"Squid are the jocks of the cephalopod world. They swim very fast. They're designed for speed," he said. "The octopus is the intellectual. They can solve problems and learn quite remarkable things."

We may share many things with squid—a similar eye, a similar neuron, neurotransmitters like dopamine, perhaps even certain intellectual proclivities—but there is one biological area in which our styles decidedly differ: sex.

HEURE D'AMOUR

Amoebas at the start
Were not complex;
They tore themselves apart
And started Sex.

—ARTHUR GUITERMAN, POET

I t wasn't until live worms—long and white and writhing—
emerged from the carcass of the two-years-dead *Dosidicus
gigas* that shrieks of joyful terror filled the Newburyport
High School dissecting lab.

"What's this?" one student asked commercial-fisherman-
turned-teacher Rob Yeomans. He was holding a white mass in
his hands. It was about six months before Rob met me to go
Dosidicus fishing on the West Coast, and he was helping his
marine biology students dig into the innards of a carcass sent
to him by Bill Gilly. I was up there, too, watching, along with
several other adults.

When Rob heard the shrills, he turned to look. Already the
things had managed to spread. At about three inches in length,
they looked like short tapeworms. The lab table was covered
with these squirrelly, hopping things that seemed to be able
to stand up on their tips and cavort along the metallic surface,
like weird animation creatures. One adult decided it was as
though you'd been sitting at a bar and had one too many and
the bartender's cocktail straws started crawling toward you like
inchworms and then stood up and danced. To me, they looked
like huge, distorted jumping beans.

The unofficial consensus was that they were some kind of
unidentified parasites that had stowed away in the squid and
managed to survive two years in deep freeze. Then, like something
from a B-movie horror flick, they'd reawakened on the other side
of the American continent in this high school dissecting lab.

Before starting to carve up the carcass, Rob had carefully instructed the kids to put on latex gloves. Most complied. One boy, though, declined. He wasn't worried. His dad was a scientist. He was experienced at this sort of thing. When the jumping started, he'd been holding a pile of the white things in his hands.

A student holding the mass of "worms"

Warmed by his palms, the mass came to life and began squirming in his hands. The boy's face turned ashen. He put the mess down, headed over to the faucet, picked up the soap, and began scrubbing. Minutes later, he was still scrubbing. Whatever the things were, he didn't want them burrowing into his skin. Which, in reality, is what they might have done if he'd held them long enough.

The kids could thank Rob's intrepid curiosity for the presence of *Dosidicus* that day. Having seen several television programs about Humboldt squid that featured the work of Gilly's lab, Rob

Yeomans helping to dissect in classroom

had poked around on the Internet, found the lab's Web site, and e-mailed, asking for a Humboldt carcass for his students to dissect. Rob became the first schoolteacher to take official advantage of Gilly's Squids4Kids educational project. This dissection was the result of that request.

You wouldn't think that a biology lab class would be The Main Event in a high school, but all day long kids not in Rob's marine biology class had begged him to be allowed to attend. Their excitement was fueled by the long string of docudramas on a variety of cable stations about how dangerous and horrific *Dosidicus* is. Rob had had to adopt a very stern voice—"No! No, *please!*"—to keep the class from being mobbed by party crashers. By the time the late afternoon class began, Rob had had to recruit his department head, a physics teacher, to guard the door.

The dissection started out smoothly enough. Several boys lifted the thawed carcass out of its container and put it on the lab table. Then a line of girls elbowed their way in to form a phalanx at the dissecting table. They looked like groupies in a mosh pit. There was no room in the front line for the boys, who stood behind and watched, arms folded across their chests.

The girls reveled in the yin and yang of their loathing and fascination. The frenzy mounted. Various bits and pieces of anatomy were pulled out—the stomach (it turned out there was a whole, not-quite-thawed, not-yet-digested fish inside), the heart, the gladius (a.k.a. the pen), the eyes, the ink sac (ink flowed out), and various bits and pieces of brains. One girl spent most of her time in a trancelike state picking the sharp little rings out of the squid's suckers. She was deeply intent on trying to harvest as many of the toothed rings as possible. Later that day she went home and shocked her mother by saying she wanted to switch her career goal from baking to marine science.

After pulling the beak out and cleaning it off, the kids passed the thing around in triumph, like a war souvenir. Only a few students stood far to the back, squeamish. Rob didn't force them to come up and participate, as long as they took good notes. It may have been the most intently focused high school lab dissecting class in the history of Newburyport High School.

But when the translucent white things began to shiver and quiver and crawl and jump, even Rob was taken aback.

"They really are moving," he said.

As the master of ceremonies, he wasn't quite as calm as before.

"This thing has been frozen for two years, and there's something moving inside it . . . ," he said, as though talking to himself.

Some kids talked about stuffing the jumping objects into the storage closet.

"Come in tomorrow, and open the storage closet, and out they'll pour," one said. These kids had seen a lot of horror movies.

· : · : · : · : ·

The "jumping beans" turned out to be packages called spermatophores. Spermatophores are densely packed semitranslucent capsules that become filled with sperm. Spermatophores embed themselves into the flesh of female squid, then bide their time. They are also capable, apparently, of dancing on metal when the occasion calls for it. At some point, with enough oxygen and under the right circumstances (scientists are not entirely sure what those circumstances are) the capsules explode. The sperm is finally free to ooze wherever it can.

Most cephalopods use spermatophores in reproduction, but the strategy as to how to use them varies, sometimes considerably, from species to species. Spermatophores from *Dosidicus* have "tabs" on their ends, proteins that are capable of becoming chemically active and agitated when the time is right. Apparently, during Rob's classroom dissection, the time was right. Hence the seemingly suddenly alive "things" with the ability to wiggle and jiggle.

Squid sex is not a pretty sight. It's not even a salacious, titillating, or scintillating sight. In fact, most of us probably wouldn't recognize sex in squid for what it is even if we were fortunate enough to see it. Of course, that's my opinion. Not everyone agrees. Gilly said: "It's not that different from humans, except that they use their hands." Then, laying a finger beside his nose, he paused. He thought a bit: "Maybe it's more like artificial insemination."

Squid and humans may share a similar neuronal design, but when it comes to sex, we share only the most basic sperm-meets-egg stuff. The physical act itself is quite different. And the more I learned about squid sex, the more I thought that our differences might be a good thing. For humans, at least.

In most species of squid, the male has a special spoonlike tool, called a hectocotylus, on the end of one of its arms. With this tool, the male takes some of its own stored spermatophores and places them somewhere on the female's body. Where the male places his sperm seems to be species-specific.

California market squid, the kind caught in Monterey Bay and sold to China and Japan, is the West Coast equivalent of *Loligo pealei*. Because it's a commercially valuable species and its numbers seem to be declining, we've spent a lot of money learning about its lifestyle preferences. The male usually places the sperm inside the female squid's mantle. The idea is that when the female expels her eggs, some of those eggs might brush past the male's spermatophores and the male might get lucky. His genes get to wiggle their way into the next generation.

This doesn't seem to me to be a very satisfactory arrangement for the male. Even if he does the dirty deed and succeeds in stuffing the sperm somewhere on the female's body, he has no guarantee that it's his sperm that will win the ultimate race to create the next generation. As far as science knows right now, there's a strong element of chance in the male squid's approach. Of course, some observers might say that's true for male humans as well.

In general, even in market squid, we know very little of the details of squid sex, but which sperm are successful appears to be at least in part a matter of accident. Female choice seems to have little to do with the final product, although there may be a way in which the female squid exercises her judgment that we don't recognize. As any guy knows, female choice is not always easy to discern. Rather early on we understood that the male peacock's tail feathers were the lure for the peahen, who chooses which male will finally get to fertilize her eggs based on the flamboyance of the male's tail. But female decision-making isn't always that obvious. Evolutionarily speaking, it doesn't make sense that success for male squid is a mere matter of chance. There ought to be some kind of fitness test that helps the female discern which sperm is best, some way for the male to prove that he's stronger or smarter or a better swimmer than the other guy, so that the female will choose him.

Market squid do exhibit at least one male sex behavior that has also been seen in male humans: Some male market squid become "guards." That is, after they deposit their sperm in the female's body, males may expend a good deal of effort in trying to keep the other guys away. I talked to market squid expert and Gilly colleague Lou Zeidberg about this, and his description reminded me of Joe DeGiorgis's "bar scene" description of mating male *Loligo pealei*.

"Guard males" are not always successful. Zeidberg has seen "sneaker male" mating in his market squid, both while watching market squid at sea and by studying many incidents recorded by undersea video equipment. He's seen that the larger males stick around to ward off other males. But he's also seen the little guy make his own opportunities. While the big males fight it out, a little male might zip in and stash his stuff.

So it turns out that the wimpy kid is not so wimpy after all. Zeidberg's colleague Miriam Goldstein calls the sneaker-male tactics "drive-by sperming." Using this strategy, the smaller male deposits its sperm in a pocket near the female's beak area, which turns out to be a location the bigger guy isn't paying that much attention to, since he's spending much of his time trying to keep the other bigger guys away from the female's mantle. The female stores that sneaker sperm until she's ready. This storage may occur for a period of hours, days, or even longer, Zeidberg suspects. When the time comes, the female market squid exudes eggs from her oviduct, which are then fertilized by sperm from the guard male. Then she holds those eggs in her crown of arms and stuffs them into an egg sac. This act occurs right near where the sneaker male left his sperm. That, said Zeidberg, is probably how at least some of the sperm from the smaller animals gets into the next generation of market squid.

"We've done paternity tests," Zeidberg said.

At least in California market squid. Zeidberg and his colleagues found out that roughly 80 percent of a female's eggs are fertilized with sperm from the guard male, but about 20 percent or so are fertilized via the drive-by strategy. And so science has proven once and for all that the wimpy kid does get the girls.

I asked Zeidberg why a species would adopt two different strategies.

"It's sort of like hedging your bets," he said. "You've got two different life strategies. One is really good for most of the time." But if something changes in the ecology of the ocean, the other style of mating may provide some important kind of backup to the species population as a whole. Why the other strategy for squid mating might one day be needed, Zeidberg didn't know, but he explained how a similar situation—two lifestyles in one salmon species—allowed for species survival. Most salmon swim out to sea, he explained, but a small portion may stay in nearshore waters around the mouth of the river where they spawned. If something at sea—some kind of ecological change, perhaps, or a mass of floating toxic plastic—wipes out the salmon, there will still be this small group of less adventurous salmon left to start up a new population.

So, variety really is the spice of life. In keeping with that philosophy, there exist many species-dependent variations of squid sex. The eccentric, semi-preternatural, bioluminescent, eight-armed, deep-sea *Taningia danae* may weigh several hundred pounds and perhaps reach twice the size of Julie's squid. *Taningia*'s strategy for love makes drive-by sperming look downright gentlemanly. This species is rarely seen, but Dutch scientist Hendrick Jan Hoving has studied several female carcasses. He found a number of flesh wounds containing spermatophores in a variety of places on the females' bodies.

He wrote that he found said wounds "suggestive." Apparently, the male *Taningia* uses its sharp beak to slash the female's flesh. The animal then inserts the sperm into the wound. Is this love?

Architeuthis is not a whole lot kinder. The Gilly lab's Danna Staaf, a passionate squid sex aficionado, explained: "We think giant squid males have these massive muscular penises and they rub them up against the female with these powerful arms and inject ropes of sperm under their skin. Of course, that's never been observed. It's only a theory." The technique, which has been observed in other squid species, is called hypodermic reproduction. Danna declined to comment on whether the female squid enjoys this approach.

Then there's the paper nautilus, a.k.a. argonaut, actually an octopus rather than a squid or genuine nautilus. "What's so cool about the argonaut," said Danna, "is that the males are about a tenth the size of the females. And there's this special arm, a hectocotylus, that the male has for passing spermatophores. The arm in the male is the same size as the male's whole body, or even bigger, maybe even twice the size, and when it's time to mate the arm breaks off and swims to the female, holding the spermatophores. For years, people looked for the male of the species and couldn't find it, but they found these females and thought there were big worms inside them, and they named the worms 'hectocotylus.' So that's how the hectocotylus got its name. Only then they found out it wasn't a worm but part of the male, so instead of meaning a parasitic worm, the word came to mean a specialized arm." Truth to tell, there may be no one on the planet who knows more about squid sex than Danna Staaf.

Little is known about the physical act of sex in *Dosidicus,* although in May 2004, in the Sea of Cortez near the southern tip of a group of small islands, Gilly witnessed twice what he thought might have been a mating pair of *Dosidicus.* He and a group of colleagues were standing on the deck of the research boat on a very still night when the water was dead calm. Floating motionless right near the boat he saw a large squid, which he thought was probably the female. The tips of its fins were above the water's surface. "In its arms, it was holding the smaller squid, which if they were mating was hopefully a male," he said. "There was no struggling. It didn't look like the larger one was trying to eat the smaller one. I could see, flickering near the side of the big squid's arms, the tip of the smaller squid's arm coming out, stroking, tickling the female's arm."

The animals hung there together for about ten or fifteen minutes, then just passively sank beneath the surface.

"It's not something I'd seen before," he said. "I'd seen squid at the surface before, but not two of them, holding each other."

"Were they mating?" I asked.

"I don't know what else they would have been doing. It's sort of a romantic notion, to think they were mating. I suppose I could have captured them and looked for eggs and sperm, but it wouldn't have been a very popular thing to have done on that boat. We were outnumbered by women at the time. It was very beautiful, so I think it was best at that point just not to have discovered too much."

In the case of squid, the eons have allowed for the evolution of all sorts of ways to enjoin sperm with egg. As the scientists say, in the ocean there are whole libraries of genes with almost limitless strategies.

· : · : · : · : ·

We, of course, have our own strategies for sneaker males. In nineteenth-century France, for instance, upper-class gentlemen, who regarded themselves as quite civilized in matters of female choice, respected the custom of *heure d'amour,* the hour just before dinner when a husband would be wise not to visit his wife's boudoir. Squid and most other cephalopods have their own *heure d'amour,* but combat for the ladies' attention may sometimes be anything but civilized. In some cephalopod species, the fight for the female is sometimes to the death. But even more striking is the fact that even when the males win, they die anyway. So do the females. They release their eggs and drift off into the Big Sleep. Cephalopods in general get one, and only one, *heure d'amour.* After the act's over, it's all over for the squid.

This is another reason why it doesn't bother me much that we don't have much in common with squid when it comes to the physical act of love.

· : · · : · : ·

Humboldt reproduction was also a point of interest on Julie's research cruise. By the time darkness fell on the research boat that November evening in Monterey Bay, the crew had pulled in a good many specimens. *Dosidicus* tentacles slid everywhere over the boat's deck. The place seemed to writhe with snakes. This was no place for Indiana Jones. When Gilly pulled in one of the animals and dropped it on the deck, a tentacle immediately began slithering over his boot and up his leg. (If you go *Dosidicus* fishing, it's probably best not to wear shorts and sandals.) After about sixty or so animals had been pulled up, the crew was tiring. It's no easy thing to reel in animals that size from hundreds of feet below.

Gilly placed the last of the whole squid in plastic bags in the cooler for use in his Squids4Kids program, and Julie began to slice and dice the remaining squid for research purposes.

Holding the flesh of the mantle apart with her right hand, she scooped the spermatophores out of one squid's body with her left.

"This one's a male," she told Gilly.

"Save some of the mantles," he answered. Voicing his epicurean side, he added: "I want to smoke some again."

Behind her, Rob officially pronounced himself done, after lugging his last *Dosidicus* on board.

"Packing it in, Rob?" Gilly asked. "You don't have to quit just because I did."

Julie turned around in surprise as Rob's last catch let loose from the siphon and she was showered with cold seawater. This made his total for the night around twenty, and even the effervescent Rob was exhausted.

Julie began processing stomach after stomach, cutting them out of the animals' innards and putting them into plastic bags. In a few days, she would head over to John Field's government lab to look at these samples again.

CHAPTER TEN

After a respite of about thirty seconds, Rob headed over to help Julie with her task and to get an advanced lesson in *Dosidicus* dissection. Julie and Rob checked the female carcasses to see if they'd been mated. In the case of Humboldt squid, the spermatophores embed themselves into the flesh around the female's mouth. Then they work their way into her flesh. They remain there, looking like fairly innocuous pimples, until the female is ready to spawn.

Julie wanted to gather data on how many of the squid they'd pulled in that night had been mated. She and Rob examined the flesh carefully around the animals' mouths, and sure enough, she was able to point out to Rob several "pimples" around some of the females' mouths.

Julie pulled out more spermatophores to show him. They were, as they had been in Rob's high school classroom months earlier, long and translucent, white and plentiful and restless in her hand. Covered with some kind of mucuslike material, they slid over her palm and dropped onto the boat deck.

· : · : · : · : ·

Sex was not initially a part of the animal world. The very first species of animals that evolved used techniques to reproduce that did not involve the ritual interactions of males and females. Amoebas, for example, reproduce by just dividing. In theory, this is a better approach in some ways than what we've got. It certainly requires a lot less emotional angst.

Nevertheless, around a billion years ago, give or take a few million years, the idea of male and female took shape on planet Earth. No one can really explain scientifically why this happened, although a lot of intriguing explanations have been proposed. Chris Adami of the California Institute of Technology, who uses computer programming to study genetic recombination, suggested not so long ago that life on our happy Eden, planet Earth, was disrupted not by an apple, but by some kind of mass disaster like a large meteor impact. Genes could have been juggled around. Such a disaster could have caused a higher rate of mutation in our amoeba-like ancestors and led, eventually, to the solution of genetic recombination and male and female.

Whether or not this was a good idea, in the human species at least, remains to be seen.

· ∶ · ∶ · ∶ ·

A few weeks after the cruise, Julie headed over to John Field's research lab in Santa Cruz, on the northern coast of Monterey Bay. She was there to study the stomach contents removed from the squid caught and dissected during the evening research cruise. They were trying to find out about what squid of various sizes ate during various seasons, a task Field had been at for years. Julie hoped that some of the data would provide information for her thesis. The next hours would be filled with a variety of tasks that would wind up filling Julie and John Field's master files with endless columns of numbers that could result in a solid theory as to why *Dosidicus* had so suddenly appeared in such great numbers.

Field had taken the frozen stomachs out of the freezer a day earlier. The team needed to weigh the stomachs with everything still inside, then take all the stomach contents out and weigh just the stomach lining. This would also provide the figure of how much the contents weighed.

Then came a task something like "panning for gold," as Julie described it. She placed the thawed stomach contents in a sieve and placed the sieve under running water. "This washes most of the small stuff away, and you're left with the bigger stuff—eye lenses, bones, otoliths. You find parasites, shells from things they've eaten, and there's always some stomach junk—unidentifiable juices, stuff like that."

Each piece of detritus must be pulled out and identified. The otoliths—fish organs similar to squid statoliths—must be pulled out and cleaned off. Otoliths (and statoliths) are unique to each species, so scientists can use them to identify what kinds of fish a squid might be eating. Some otoliths may be easily visible, but others are quite tiny, perhaps because they come from a young fish, or perhaps because they come from a species of fish in which otoliths are routinely small. Each otolith must be identified and documented. There are scientists who have devoted their entire careers to identifying and cataloguing various otoliths and statoliths. Field has thick identification books beside him at his microscope in the lab, in case he comes across an otolith he can't identify. That doesn't happen often, though, since after years of studying squid stomach contents, he has become somewhat of an expert in his own right. Every once in a while, when he can't make an identification even by using the catalogues, he might pack up a strange otolith and mail it to an expert for a final ID. Examining the contents of each individual stomach requires anywhere from twenty-five minutes to an hour.

I asked Julie if she minded all the tedious work. With her usual ebullience, she said no: "It's like mindless work, poking around, but it's nice to have an afternoon like that, sitting and being engaged the whole time. . . ."

"Zenlike," I commented. She agreed.

When Julie turned up in Field's lab that day, she was particularly excited about a result from the research cruise. The tracking tag on the *Dosidicus* she had so gently slipped back into Monterey Bay had finally turned up. It had come off the animal in seventeen days, just as it had been programmed to do. The tag's satellite system showed it to be about 100 miles offshore of Ensenada, Mexico.

"Cool," was her response. After she figured in both horizontal and vertical migration distance (the squid goes up and down each day, making about a mile round-trip), she averaged this specimen's travel to about 35 kilometers or 22 miles a day— almost marathon distance.

The finding was important, providing the factual information to back up a scientific theory about Humboldt squid migration. Gilly had seen similar migration patterns, but only over the course of a few days at any one time. Julie's tagged *Dosidicus* confirmed for the first time that these squid were capable of sustainable perseverance; that they could travel considerable distances over a fairly long period of time.

"That they could do that over seventeen days is pretty impressive," she said.

Field said: "We kind of thought that they might have moved up and gone back, but now we have proof."

Field has been paying attention to the migration patterns and feeding behavior of *Dosidicus* for the better part of a decade. There are records of *Dosidicus* being present in Monterey Bay in the 1930s, but apparently the population didn't stick around very long. "The 1997–98 El Niño resulted in an unusual persistence of the new population," he wrote in an important overview paper on the mystery of the animals' sudden proliferation.

Over the first decade of the new century, *Dosidicus* populations seem to be spreading along the west coast of both North and South America. The upside is that this area is now one of the world's largest cephalopod fisheries. The downside could be that the explosive populations are gobbling up everything they can find to eat, including commercially valuable rockfish and hake. Their large populations may not be a good sign, as far as the ocean's ecological health is concerned.

Dosidicus is very "opportunistic," Field said. "They seem to do well under disturbances. There are some studies that suggest that in heavily fished ecosystems, cephalopods seem to do very well. They're well-suited to take advantage of changing conditions, since they have short life spans, high growth rates, and very high potential fecundity."

His description reminded me of a study I'd read years earlier about coyotes. It turned out that the more ranchers and farmers tried to get rid of the animals, the more fecund the coyotes became. If a female coyote had a territory that was fairly open, she had more pups. If her territory was rather full, her litter would be much smaller. With their very flexible behavior, coyotes are happy on exclusive golf courses and in wilderness areas, and on Cape Cod they are well known to enjoy a good meal of mice or of watermelon gleaned from Dumpsters after the Fourth of July.

I wondered if cephalopods were generalists, like coyotes. Could that be one reason for their existence over hundreds of millions of years?

Field and I chatted a bit about the paleological record. Cephalopods made it through several major extinctions, including the 250-million-year-old Permian or "Great Dying" extinction in which about 95 percent of living species disappeared.

Is it any wonder, I asked him, that *Dosidicus* seems to be thriving right now? After all, jellyfish—very ancient and primitive animals lacking brains—are also suddenly spreading worldwide.

He said he wasn't surprised.

"Squid outbreaks like this can persist for a while," he answered, "or they can disappear, or they can stay around for a very long time. They're probably going to be around in this ecosystem now for quite a long time. If the ocean is changing as much as we think it is, they're probably going to be around along the California Current for the long term."

PLAYDATE

In their brief time together, Slothrop formed the opinion that this octopus was not in good mental health.

—THOMAS PYNCHON

C an an animal with its brains wrapped around its throat really be smart? It's an intriguing question, isn't it? Thousands of years ago, Aristotle pronounced the octopus "stupid." That view prevailed throughout much of Western civilization. Until very recently.

· : · : · : ·

On an early June day, 2009, while the sad citizens of Boston were still wrapped in coats and huddled against the depressingly endless cold rain of that particular year, Wilson P. Menashi, in shirtsleeves, was in a back room at the New England Aquarium. He was standing in a rather large puddle of water, mulling over this and other questions of identity and intelligence and the ultimate meaning of life.

Menashi was seventy-five. His companion in meditation, Truman, was about two. Both were drawing on their individual life experiences, both having long passed the halfway point of their individual life cycles, which in Truman's case would probably be just a bit more than three years.

Menashi had decades earlier helped invent cubic zirconia. Having retired early, he had for the past fifteen years volunteered one day a week at the aquarium, acquiring in the process the status of a sort-of staff member, albeit an unpaid one. His daughter had pushed him into it, to keep him from hanging around the house when he stopped working at Arthur D. Little. But no one had needed to make him keep coming back. He loved the place.

Truman, on the other hand, had come here not of his own free will—if, that is, a giant Pacific octopus (*Octopus dofleini*) can be said to have "free will." Or any will at all, for that matter. Whether Truman loved the place or not, no one could say. He certainly seemed to "love" Wilson, though. Or, to make some scientists more comfortable, we can with certainty say that the animal was undeniably drawn to the man. Wilson suspects, with a degree of modesty, that he and Truman have a special relationship, a thing going on.

"My feeling is that he'll let me do things with him that he might not allow others to do. Of course, I don't know that that's true. It's just my feeling," Wilson told me.

As I walked through the widening puddle over to the pair to introduce myself, Truman and Wilson were quietly engaged in an intimate dance, a kind of pas de dix—a dance of ten arms. Wilson's two arms gracefully clasped, as well as they could, whichever of Truman's fluid, boneless eight arms "decided" to wind around Wilson's.

Is "decided" the correct word here? Who knows? Scientists know so little about these creatures of the cold Northern Pacific that we cannot say for sure. But we do know that roughly three-fifths of an octopus's neurons reside not in the brain but in the arms. The octopus's distributed intelligence would seem to imply that the arms have "minds" of their own. (Or, at least, it

Wilson Menashi and Truman

would imply that, if we could say with any certainty what a "mind" is. . . .)

One thing is certain: Experiments have shown that a blind-folded octopus can use its arms and suckers to tell the difference between various objects, leading scientists to believe that the arm's ability to perceive by touch and by scent is as important as the animal's ability to perceive by sight. The finding isn't that surprising, since the octopus generally hunts at night, but how the various neurons interact with each other remains a mystery.

The organization of the octopus brain is strange, at least from our human point of view. The animal has a central brain with roughly 45 million or so neurons. There are two large optic lobes with about 120 million to 180 million neurons. The arms contain the rest of the roughly 500 million neurons that comprise the animal's whole nervous system. Some of the groups of neurons correspond very roughly to some of the groups that we humans have in our brains, like the hippocampus (involved in human learning and memory), for example. Some do not. We do not yet know which neurons are in control at any one time, if, indeed, a system of oversight exists at all. We do know that the suckers of a giant Pacific octopus are capable of behaving with the kind of fine-tuned dexterity with which our fingers and thumbs behave. In fact, giant Pacific octopuses may in some ways be even more dexterous. Aquarium staff who look after the octopuses have learned that they have to surround the animals' tanks with a synthetic material like Astroturf, one of the few materials the suckers cannot grip.

A few scientists have taken the first steps in trying to unravel the mystery. Hebrew University's Binyamin Hochner and his colleagues studied the movements and nerves of an octopus arm. They learned that the arm moves and reaches out using a stereotypical flow that begins near the octopus's head and body and flows out in a kind of hook that eventually reaches its tip. A flow of energy ripples down the arm. "Despite the fact that an octopus arm has virtually infinite degrees of freedom, arm movements are executed in a stereotyped manner," the authors wrote in one paper. These stereotyped movements persisted even when scientists severed the neural connections between the animal's main brain and the neural system in its arm.

And indeed, if an arm breaks off an octopus head and body, the arm often continues along its way, doing whatever it was doing before it became severed. This happens even if the octopus itself "decides" to break off one of its arms and abandon the appendage, in the midst of battle, perhaps. The arm might continue to live for several hours before finally dying.

Nor does the octopus appear to miss its appendage that much. While the severed arm goes its own way, the remains of the arm still attached to the octopus might bleed a little bit of blue blood, then regenerate a new arm, complete with nerves, flesh, and suckers. Octopuses are thus real-world, real-life versions of the virtual Namekians in Dragon Ball.

As I watched, Truman's arms began their journey upward, toward Wilson's face, by initially attaching themselves to Wilson's hands. Interspecies contact was tentative at first. Only the tiniest suckers at the end of the octopus arm contacted the man. Those suckers, with their strong muscles and high-powered chemoreceptors that act like taste buds, apparently enjoyed what they experienced, because they continued their journey of exploration, feeling their way up Wilson's arm, past the elbow joint, and over the biceps toward the man's shoulders. Meanwhile, larger and larger suckers, closer to the base of the octopus's arm, attached themselves to the man's hands and wrists.

$\cdot \ \vdots \ \cdot \ \vdots \ \cdot \ \vdots \ \cdot$

Our cultural history is filled with horrific tales of humans being captured this way by octopuses and drowned in the depths of the deep dark sea, many of which I had read by the time I made this visit. "For it struggles with him by coiling round him and it swallows him with sucker-cups and drags him asunder," the Roman Pliny the Elder wrote rather dramatically, nearly two thousand years ago.

Pliny's point of view has been held by *Homo sapiens* for much of the last two millennia. Octopuses as symbols of dangerous envelopment turn up surprisingly often in Western literature and art. In 1802, the French naturalist Pierre Denys de Montfort presented to a chapel a woodcut of a truly monstrous eight-armed, bug-eyed octopus as large as the three-masted ship it was seizing. Three octopus arms entwined the ship's masts like snakes. Two other arms grasped each end of the ship, as though to pull the ship closer to its beak. The other three arms hung there, unoccupied, as though perhaps waiting for sailors to fall into the sea and be eaten.

de Montfort's woodcut

Although scientists generally agree that, for reasons unknown, many species of sea life grow to much smaller sizes today than in the past, the likelihood of an octopus (or any invertebrate) growing to such a size appears doubtful. Not to everyone, though. Shedd Aquarium biologist Roger Klocek has written that de Montfort's woodcut and other evidence have led him "to believe an octopus with an arm span of more than 150 feet did exist" at one time.

Maybe. But I think it's more a symptom of our own special fears. Evolutionary theory speculates that life first emerged from the salt water onto the land as a defensive measure because it was just too darned dangerous in the world's oceans. That's a strategy that makes sense to me, but still, I'm doubtful that octopuses were ever large enough to attack ships. It's also quite possible that early observers confused the octopus with the squid. A giant squid swimming in the water or floating on the sea surface can look quite like an octopus and, if you're not prepared, be a frightening sight. But even giant or colossal squid are unlikely ever to have been that large.

After de Montfort's woodcut was presented, French culture continued to be obsessed with dangerous cephalopods. In the nineteenth century, an octopus—and not a very nice one, either—became one of the main characters in Victor Hugo's *Toilers of the Sea*. "Suddenly he felt something seizing hold of his arm. He was struck with indescribable horror," wrote Hugo about his protagonist swimming off the French coast. Ultimately, Hugo's octopus tries but fails to drown the novel's hero. The animal does, however, ultimately succeed in dragging the novel's villain into its lair, where only his skeleton will later be found.

Paris loved the malevolence of Hugo's octopus. Soon after the novel's publication, Parisian women began wearing hats with octopus arms hanging from the brims. "Everything is octopusied," exclaimed one contemporary French letter-writer, commenting on the new fashion.

Of course, it wasn't only the French who were obsessed with the fearsomeness of octopuses. Gerald Durrell, the renowned British twentieth-century natural history writer, who was otherwise effusively in love with all things natural, compared an octopus sitting on a rock to a "Medusa head." And American author Frank Norris chose *Octopus* for the main title of his muckraking book that described how, before Theodore Roosevelt's presidency, railroad corporations entwined prairie farmers in an inescapable economic stranglehold. Even nature-oriented John Steinbeck, in *Cannery Row,* depicts the octopus as a "creeping murderer" that stalks its prey, "pretending now to be a bit of weed, now a rock, now a lump of decaying meat while its evil goat eyes watch coldly." Thomas Pynchon's *Gravity's Rainbow* features a huge octopus that wraps an arm around a woman and tries to drag her into the water. Pynchon's hero, Slothrop, saves her by beating the octopus over the head with a wine bottle. Writes Pynchon: "In their brief time together, Slothrop formed the impression that this octopus was not in good mental health."

Perhaps in response to the belief that octopuses like to carry women into watery graves, twentieth-century American manliness was for a time expressed by willingness to wrestle an unsuspecting giant Pacific octopus out of its den and drag it up to the surface and onto land, where it would, eventually but inevitably, die. Octopus wrestling seemed to be a symbol of some kind of manly American bravery. "When the native lunged for the purplish eyes of the giant octopus, the monster caught him with one of its writhing tentacles," wrote Wilmon Menard in "Octopus Wrestling Is My Hobby," a 1949 story in *Modern Mechanix.*

Modern Mechanix illustration of octopus drowning person

In this story, Menard as Caucasian male hero saves the hapless native by wrestling the octopus into submission.

In fact, oddly, octopus wrestling was a very big deal in other places as well. In Seattle, until the late 1960s, octopus-wrestling contests were thought to be the true measure of manliness among the Lloyd Bridges subset of scuba divers.

William Beebe, of course, found octopuses repulsive. "I have always a struggle before I can make my hands do their duty and seize a tentacle." But not all writers have felt this kind of distaste. In the most elegant paeon to the octopus I've ever read, the naturalist and journalist Gilbert Klingel wrote about a quite large specimen he encountered when marooned on the southern Bahamian island of Managua at just around the same time that William Beebe was descending to the ocean depths in Bermuda.

"I feel about octopuses—as Mark Twain did about the devil—that someone should undertake their rehabilitation," Klingel began his graceful essay, "In Defense of Octopuses." He complained about octopus wrestling, writing that "no one has ever told the octopuses' side of the story." He told his own tale of horror, of when he encountered a large octopus in Bahamian waters that he thought at first was an orange rock, which he intended to use as a handhold. "Before my gaze, the rock started to melt, began to ooze at the sides like a candle that had become too hot." But he overcame his fear and began to observe octopuses in the water and research them in the literature.

"I have found among them animals of unusual attainments and they should be ranked among the most remarkable denizens of the sea," he continued. "Had they been able to pass the barrier of the edge of the ocean as the early fish-derived amphibians did there might have been no limit to the amazing forms which would have peopled the earth."

This kind of positive PR is rare. Most of our popular literature and art vilifies the animals. I must confess that I myself was not overly fond of the octopus when I started writing this book, although I couldn't explain why. Then I found the answer, which dates back to my childhood. To begin researching this book, I collected a lot of old films and began watching them. Among those was *It Came from Beneath the Sea*, a 1955 B movie about a giant octopus that, fed up with humanity, began wrapping its anatomically incorrect six arms around women and children. The octopus, about as big as the Transamerica Pyramid, dragged victims—including little girls—off the streets of San Francisco into the waters underneath the Golden Gate Bridge. There, presumably, the victims died unhappy deaths. As I watched the film a second time, I realized that I had seen this movie as a child, and that it probably explained some of my fear of sea monsters.

On this particular chilly June day in Boston, however, Wilson Menashi didn't appear a bit fearful. The rhythmic embraces of the giant Pacific octopus seemed almost to soothe him. Wilson could have been playing quietly with his pet dog. The puddle he was standing in was getting wider by the minute. But as it was only an inch deep, death by drowning did not seem imminent. On the other hand, who knew what this kind of intimacy could lead to? That was my way of thinking, anyway.

"You have to keep playing with them," Wilson said, casually peeling suckers off his neck and shoulder and throwing an arm gently back into the octopus tank. "They get bored very easily. They simply enjoy doing things. They'll work on a puzzle for a long time. They do lots of things just because they want to."

Or they won't want to. Wilson, wearing his creative engineering hat, has devised all kinds of puzzle boxes for his various giant Pacific octopus clients, and he's found over the years that some of the animals will stick to the task of solving a problem and some will give up and go into a corner or behind a rock. Others are simply not interested at all.

Mark Rehling, an aquarist with Cleveland Metroparks, believes that the ability to persevere is partly a factor of age. Older octopuses seem to be able to focus better. But he also believes that this ability has something to do with individual personality in each animal. This concept that an octopus would have a personality is fairly new in the field. Rehling created all kinds of complicated objects, which he calls "prey puzzles," objects containing food that require problem solving. In general, the older animals were more successful, indicating that some kind of learning process, some change in neural connections, occurs. Hebrew University's Binyamin Hochner also maintains that there indeed are changes in an octopus brain in regards to learning and memory that are, in some ways, quite like the changes that occur in our brains when we learn and remember things.

Rehling has found that a few adult octopuses even refused to give back the puzzle pieces once they'd eaten the food hidden inside. Some animals had more difficulty than others, which implies some level of personality, intelligence, and problem solving. One thing was clear, Rehling writes: Prey puzzles originally designed for primates were just too "simplistic . . . if the octopus was interested, a solution was sure to follow."

Truman is one of the dedicated puzzlers. Only weeks before I met him, he had achieved international fame and quite a bit of glory because of this interest. Bill Murphy, the aquarium's head aquarist in the octopus division, had put a live crab into one of Wilson's box-within-a-box puzzles and given it to Truman. Since giant Pacific octopuses tend to hunt at night and lie low for the day, Murphy expected Truman to do what he usually did: envelop the box with his whole body and carry it away into hiding, to work on at night when the lights were low.

Not this time. In full view of an astounded public—at least one of whom had a video camera—standing on the other side of the thick tank glass, Truman began the same testing process he would later use on Wilson's arms. First, the tiniest tip of an arm entered the large outer box through an extremely small opening of only a few square inches. Soon, the whole animal, all 30 pounds and eight arms and innards-filled mantle, had oozed its way into the 15"-by-15" clear plastic outer box.

Now Truman was almost flattened pancakelike as he squeezed inside the outer box, but outside the inner box. How did he make the decision to behave this way? It's a mystery to us vertebrates, because we have a central nervous system mainly housed in a cranium. This is the part of us that we believe is mostly in charge. Whether that's true or not is a subject of intense scientific and philosophical debate, but at the very least, we have the illusion that our head is in charge of our body's behavior.

If a doctor taps our kneecap with a little rubber hammer, we usually respond reflexively. That's because there are some neural cell bodies that reside not in the brain, but in the spine. Since the message doesn't have to travel all the way to the brain, the axons are not as long and the knee jerk happens quite quickly. But we don't *have* to jerk our knee. Most of us can develop the ability to feel that knock but decide mentally to override the immediate physical response. It takes time to learn, though, since the brain must practice overriding neurons with cell bodies located so far away from our cranium.

Truman's brain is much less centralized. An octopus may have perhaps 500 million, or half a billion, neurons, as compared to a human's estimated 100 billion. The number of octopus nerve cells is just a bit less than the number of nerve cells in a dog, about 600 million, but only about half those of a cat, about 1,000 million. But sheer numbers don't necessarily correlate with intelligence. Organization of the neurons is also important. The neurons involved in our reflex actions are important, but we certainly wouldn't call the resulting knee jerk "intelligent."

Truman does have a central brain, wrapped around his esophagus. But it contains only a third of the nerve cells that process his decisions and actions. This fact greatly interests people who work on robotics and who want to know more about distributed intelligence. Where are Truman's decisions made? Does he have an apical decision making structure, with one neural region in charge and capable of overriding the others? Or does one arm "argue" with the other about how to respond in a crisis?

· : · : · : ·

In deciphering Wilson's puzzle box, after he'd managed to flatten himself like a pancake, Truman's next task was to figure out how to unlock the inner box to get the crab. Although he eventually gave up and retreated without his dinner—possibly he just couldn't maneuver well in his pancakelike state—news of his exploit flashed its way across the United States, from Boston to Florida to Texas and Los Angeles. By the next day, Truman's picture was in newspapers from Britain to Singapore.

None of this fame and glory surprised Wilson in the least.

"Is Truman your favorite animal?" I asked.

Wilson nodded.

"Your favorite of all the animals in the aquarium, or of all animals, period?"

He took a while to think that one over.

"My favorite, period," he finally answered.

As Wilson and I chatted about the vagaries of fame, Truman began to get to know me better. The same kind of tentative, ticklish tentacular attachment began. First, the smallest suckers sought out my wrist. I began to feel explored and manipulated. It was a rather odd but nevertheless intriguingly benign sensation. Next came larger suckers, and then, some very large suckers. Very large, as in almost as large as my wrist. Or so it seemed.

My eyes must have widened.

When the smaller suckers reached the level of my biceps, Wilson kindly detached Truman's arms, laying them gently back in the tank. Apparently, if you start at the top with the smallest suckers and peel the octopus arm down like a banana peel, the detaching is rather simple.

Wilson warned me about getting soaked with water from Truman's siphon.

"It's OK," I said, surprised. "It's pointing the other way."

"*That* can change," he warned. Then he winked and pointed to the siphon, which was indeed changing angles as we spoke.

I stepped back.

Eventually Truman lay back in the water. He floated upside down.

"He's tired now," Wilson said. "He wants to be alone."

FAN CLUBS AND FILM STARS

The octopus looks like a silken scarf,
floating, swirling, and settling gently
as a leaf on a rock.

—JACQUES-YVES COUSTEAU

Truman is not the only giant Pacific octopus (GPO) to have a fan club. At Mystic Aquarium in Connecticut, Sammy arrived from the Canadian West Coast when he was about one year old, just a little before the time when Truman became famous. Electra, the aquarium's senior GPO, was nearing retirement age and a new star was needed to take her place. Even before he went on exhibit, Sammy's moods ran hot and cold. It was never clear whether being a star was the right thing for him.

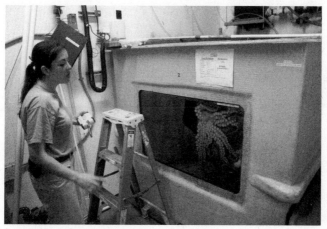

Aquarist Monique Glazier and Sammy

Sometimes he would come across the tank to interact with Monique Glazier, his young aquarist, chief provider of his daily sustenance, and all-around protector. And sometimes, he just wouldn't. Monique could never figure any rhyme or reason to his actions. "He's not always interested in new things," she told me.

"But when he is, he's *really* interested." When Sammy was moved from a behind-the-scenes role onto exhibit, the first thing he did was hide in a bunch of plastic kelp fronds. His camouflage skills were so well-honed that he just disappeared. Aquarium staff put their fingers in the water, trying to lure him out. "As soon as I put *my* hand in the water, Sammy came over to see me," Monique said. "It appeared as though he came over specifically to see me. That's probably not what he was doing, but that's what it looked like to me."

Sammy rides in a basket

When Sammy made his public entrance and Electra left the stage, Mystic staff put up a sign that explained why the new GPO was so much smaller than the old one. Giant Pacific octopuses grow at an astounding rate. In the wild, they only live to the age of three or so. But despite this short life span, they grow quite large. In decades past, a few were caught that had grown to as much as 200, 300, or even, in one case, 400 pounds. These exceptionally large specimens would be two to three times the size of a six-foot human. (There's a report from decades ago of a 600-pound GPO, but it's never been confirmed and scientists are skeptical.) If you could watch these wild animals day by day, perhaps you would be able to literally see them grow.

None as large as several hundred pounds has been found for decades. Scientists don't know why, but some speculate that the size dimin-

Senescent Electra ready to leave stage

ishment may be due to a

concomitant decrease of available food in the ocean, to greater pollution in the ocean, or to temperature changes in the ocean that have shifted the sea's various food layers. Or to all of the above. All these factors are probably swirling together to combine into an overall general degradation of the world's ocean ecosystems. It's not just the GPOs that are getting smaller. Lots of species suffer in this regard.

In aquariums, GPOs generally reach only 30 or 40 pounds, although the Seattle Aquarium had one male, Mr. Big, who weighed almost 100 pounds. Since Mystic's Sammy would start his new job as a comparatively small animal and grow quickly, some of Mystic's most faithful human fans began visiting him regularly to watch the process. Kids were especially interested, having not been raised on octopus horror stories. "There's your basic octopus," sang one ten-year-old boy on seeing Sammy. "I love cephalopods," said a young boy to his brother, adding, "There's the anemone, over there." The father was as surprised as I was by his son's vocabulary.

"How do you know those words?" I asked.

"Nemo," the boy explained. "And *SpongeBob*."

One family told me they came every week. The more faithful of Sammy's fans were invited for a personal backroom rendezvous, although Monique had to be careful, since she never knew what kind of mood Sammy would be in. He had been known to ink strangers.

Thinking that Sammy's on-again, off-again mood might perk up if he had more entertainment and less free time, Monique began designing complicated prey puzzles, just as Wilson had done for Truman. But most of what Monique created to keep Sammy busy was just too easy. Eventually Monique realized her creative juices were being hampered by her own skeleton. She was thinking "inside the box" of her own humerus, ulna, and radius.

Sammy, free of the restrictions imposed by rigid bones, could come up with all kinds of strategies for getting prey out of tight places, strategies that Monique realized she just couldn't imagine. Sammy's only limitation in changing shape was his beak. Monique hid food inside all kinds of toys, thinking Sammy would never solve the puzzle, but then there he was, back again, almost instantaneously, done with his meal and looking for something else to do. Once she put together an elaborate assemblage of pieces of colored plastic gerbil tubing, with lots of twists and turns, designed to mimic the underground tunnels that gerbils like running through. Monique imagined Sammy having to spend hours and hours of time exploring the tubing maze with his arms, looking for his meal. She imagined he'd have to elongate his arms and use his chemoreceptors to try to smell the food from a distance.

She was wrong. She said: "Sammy was smarter than I was. He learned how to break the pieces apart and get his food. Then he would feed the plastic pieces to the anemone." The first time this happened, Monique was stumped. When she fished the toy out of Sammy's tank, she knew there was a piece of plastic tubing missing, but she couldn't find it. It was too big to just disappear. And it wasn't possible (she hoped) that Sammy had eaten it. A day later, the anemone spit out the missing plastic piece. Now she counts the pieces when she puts the puzzle together.

One toy Monique created mystified her much more than it stumped Sammy. She had a baseball-size plastic ball that could be assembled or disassembled by screwing or unscrewing the two halves, kind of like a jar top. She placed a piece of food inside, screwed the two pieces together and gave Sammy the ball, which had only a few very small holes, much smaller than Sammy's beak. The animal took the ball and went away to hide and eat. The next time Monique saw the ball, the food was gone. But the ball itself was intact. Somehow Sammy had managed to get at the food hidden inside the ball without unscrewing the two halves. Either that, or he was putting the ball back together when he was done eating, but Monique was pretty sure Sammy wasn't that fastidious. This happened every time she gave Sammy the ball. "To this day, I have no idea what he does with that toy," she told me.

Monique is not the only aquarist who has trouble creating puzzles complicated enough to keep an octopus busy for any length of time. In order to help out colleagues, aquarists have created an "Octopus Enrichment Notebook," hoping that by pooling their human brains, they might be able to keep one step ahead of the octopuses' brains.

Captive octopuses all over the world have fascinated the public by being able to unscrew jars, which they do by attaching their muscular suckers to the lid and twisting. But most aquarists seem to think that this kind of problem solving is just too simple, if the goal is to keep the animal occupied for a while. One aquarist, though, did wonder why octopuses did not learn to put the top back on the jar when done eating.

· : · : · : · : ·

The first feature film ever made of an octopus, a 13-minute black-and-white by Frenchman Jean Painlevé in 1928, did nothing to rehabilitate the animal's frightening public image. The small specimen in the film, slithering over rocks on the French Atlantic coastline, looked only a little more than wormlike and was a far cry from the thrilling malevolence depicted in Victor Hugo's novel. Parisians did not mark the octopus's movie debut with hats or dresses of any kind. Its creepiness when out of the water was not overly alluring. Just watching it sent shivers up my spine.

Painlevé's second octopus film—*Les Amours de la Pieuvre* (The Loves of an Octopus)—in color this time, released in 1967, depicted the lugubrious, decidedly unexciting mating of a male and female. No titillation here: The happy couple came across as being weird without being wonderful at all. Paris, once again, yawned. The director's 1934 film of a male seahorse giving birth had been the talk of *Le Metro*—a *male* giving birth! *Mon Dieu!*—but the French public seems to have been disappointed by octopus sex. The 1967 film passed almost unnoticed from the fickle Parisian limelight.

It took Jacques-Yves Cousteau to elevate the giant Pacific octopus to the level of beloved charismatic megafauna, and to show that the octopus *in* the water was quite different from the octopus crawling over rocks. Just as Joy and George Adamson in *Born Free* showed there was a lot more to lions than predatory behavior, Cousteau almost single-handedly changed public opinion when he portrayed the giant Pacific octopus as a gentle colossus that fought only when unavoidably cornered with nowhere to hide. As a fairly young man, Cousteau had helped invent the Aqua-Lung and thus freed divers from the confines of a heavy helmet and air hose connected to the surface.

He liked to think of himself as a scientist, but there are those who would intensely disagree. Few, however, would contest the fact that his decades-long worldwide underwater filming of ocean life, most of which appeared on television around the globe, made Jacques Cousteau the best public relations agent the ocean has ever had. In fact, Cousteau has been compared to an octopus by at least one biographer, Axel Madsen: "He is tentacular, reaching out and sucking people and ideas to him."

By the time his film "Octopus, Octopus" (part of *The Undersea World of Jacques Cousteau* series) appeared in 1971, the public was already familiar with Cousteau's work and with the ocean ethic he was bestowing upon a postwar, newly open-minded, more scientifically aware world.

"A lot of people attack the sea. I make love to it," Cousteau, incomparably French, once said.

That was certainly true of his octopus film. Only the hardest of hearts would not be moved by his passionate portrait of a strange, sometimes bloblike, much-maligned animal. The film opens with a several-second shot of an apparently nonchalant giant Pacific octopus swimming calmly along underwater. Right beside the animal, a man swims just as calmly. Animal and man are about the same size, but the octopus is grace personified. When it spreads out its eight arms to glide through the ocean, it seems almost to be soaring birdlike. Who knew? Television audiences were fascinated. The octopus on land or on the surface of the sea might be frightening, but the octopus in its own subsurface environment turned out to be spellbinding.

Cousteau's point was that the octopus, completely free of the restrictions caused by a skeleton, was in the water a totally different animal than on land. "One must be able to see them slip, slide, and actually 'flow' like water to understand that the absence of a skeleton in a marine life form constitutes a form of perfection," he once wrote. In one film, he pleads the animal's cause, asking, "Octopus, octopus, are you really so unappealing?"

Cousteau's film narrator—*The Twilight Zone*'s Rod Serling, of all people, complete with his trademark cadences—begins the story by telling the audience that until recently "man could only speculate about the legendary monsters lurking beneath the sea" but that this episode would "seek the truth about this enigmatic cave-dweller . . . one of man's most curious contemporaries." For the first time ever, the public at large could watch and learn about the natural history of the octopus "beneath the concealing skin of the sea." (True to form, Serling spits out the "p" in "speculate," and the "k" in "skin," creating an aura of ominousness despite Cousteau's intentions.)

In reality, in the wild, a giant Pacific octopus is a creature unto itself. Julie's *Dosidicus* and neuroscience's *Loligo* spend their lives as members of groups, but the octopus is usually a solitary being, an animal that lives by its own wits, and one that is clearly capable of learning. Cousteau and his crew found that they could seek out the animals and habituate them, and that once the octopuses had learned to accept the presence of humans, the divers could film, for the first time, some of the routine undersea behaviors of the animals.

And when they showed these films to the world, it turned out that the octopus wasn't the malevolent being we humans had always imagined. In fact, much of the giant Pacific octopus's life in the sea is filled with what we humans would call pathos. Said to be the world's largest octopus, the animal stays very much alone throughout its life span, save for the few hours of mating which mark the beginning of the end of its time on earth.

The GPO does not establish firm territories, but instead lives in a series of temporary dens, which it may find among the rocks of the seafloor or build itself. Or a GPO might discover a suitable undersea cave, then shore it up with rocks and various debris it carries home from elsewhere. One Cousteau film showed a

Mediterranean octopus that had found and carried home a live hand grenade, presumably left over from World War II.

The octopus is a hunter, but shares only a few of its habits with mammalian predators like lions. From its temporary home, the octopus ventures short distances to forage for prey, including its favorite, crab. When it captures something, it injects its live captive with toxin that paralyzes the victim (some scientists say it also renders the victim unconscious), which it then carries home to consume. When available prey in a denning area becomes hard to find, the octopus moves on to a new home and new hunting grounds. After a period of time, if the prey has replenished, the octopus may return.

With no skeleton of any kind, the octopus is entirely vulnerable. Hundreds of millions of years ago, it gave up the protective shell that keeps its distant cousins, clams and mussels, somewhat safe. In return for that sacrifice, the octopus received the convenience of free-form movement. But unlike squid and most cuttlefish, many octopus species crawl across the seafloor on their arms probably as frequently as they swim.

Of course, as with most creatures in the world of cephalopods, there are exceptions to this general rule: Researchers recently discovered a group of octopuses that spend their entire lives swimming and may never touch the ocean floor. This group, called ctenoglossans, includes species with common names like the glass octopus and the telescope octopus.

Among most species of octopus, the first pair of arms is used less often for crawling. Instead, they seem to be aids to navigation, used somewhat the way people walking through a dark room might hold out their arms and hands to "feel" their way past obstacles. Some scientists think that this first pair of arms, stretched out, can sense the presence of prey and predators in the surrounding water. Its chemoreceptors do not need to be touching an object in order to do this. It can sense another animal through the water itself. The second pair is used for scrambling, moving over the seafloor and objects like rocks. The third and fourth pairs tend to be used for what is sometimes called "walking," although an octopus would rarely stand upright in the way that we do.

Most of the GPO's arms have about two hundred suckers, divided into two rows. The suckers are comparatively small at the arm tips and become progressively larger along the length of the arm. Near where the arms encircle the beak, the suckers can sometimes be quite large. Often described as "suction cups," the suckers are in fact much more versatile. In a certain sense, they're somewhat like fingers: Muscles attached to the inner section of each sucker allow it to operate independently of the others and to grasp or release an object. Thus, one sucker might clasp your flesh while the immediately adjacent one does not. The uppermost suckers on the arm of a very large octopus might be nearly an inch in diameter. The clasp is something you definitely notice. Unlike many squid, the GPO's suckers do not have sharp teeth or hooks. A few species of octopus have bioluminescent suckers, which some scientists think may be used as lures to draw prey near.

In regards to their arms, octopuses are capable of two behaviors that humans are likely to find rather shocking: autotomy and autophagy. In autotomy the animal separates itself from an arm, either by biting the arm off, or via a biological process that begins internally, below the skin of the arm, ending in the arm seemingly severing itself. Autophagy is even stranger. Several days before the event occurs, an arm begins to develop a kind of tremor. Eventually, the animal begins to eat its own arm. Scientists do not know why this occurs, although some suspect it may have to do with an infection, as many of the animals seen doing this have been well-fed and are not hungry.

A giant Pacific octopus may hunt as many as six times in a 24-hour period on a round-the-clock basis, but seems to prefer nighttime excursions, which scientists believe tend to be longer than daytime hunting trips. A web, a not-very-thick film of flesh that spreads between the eight arms somewhat like a skirt, extends perhaps as much as a quarter of the length of an arm and is used to envelop the prey and help bring it to the beak.

Instead of having a multichambered heart like we do, the octopus has three different hearts. There is a small heart at the base of each gill, as well as a main heart. The blood, which is blue when carrying oxygen rather than red (owing to the presence of copper rather than iron), circulates through all three hearts as the amount of oxygen is raised and lowered. When the blood is depleted of oxygen, it is almost translucent. Oxygen is a major problem for a big guy like a GPO. The air we breathe has plenty of oxygen, but the water in which a GPO "breathes" may carry much less than 1 percent oxygen. This is why the GPO seeks colder water—warmer water carries even less oxygen. And perhaps because its system relies on copper rather than iron, it seems to have less stamina. This may be why Truman eventually tired of being held in Wilson's arms.

Few animals have as heart-wrenching an end as the female giant Pacific octopus. Biologist and GPO expert Jim Cosgrove entitles his public talks "No Mother Could Give More." The female has one chance, and one chance only, to send her genes into the next generation. When she becomes sexually ready, she begins to attract males. No one is quite sure how she does this, but the suspicion is that she sends out pheromones through the water that the male "smells" via his suckers' chemoreceptors. The male mates with the female by using a special tip, called a ligula, attached to his third right arm. After receiving the male's sperm, encased in spermatophores that sometimes may be almost three feet long, the female will nurture the eggs inside her body for five months, or perhaps a bit longer, depending on how cold the water is.

When the time comes, in the cave she has chosen, she expels each egg, one by one, then—using her suckers—painstakingly *braids* them together into long chains that look like plaits of a woman's hair. These plaited chains of eggs she will then attach to the roof of her den. The whole process, which may involve exuding and braiding not quite 100,000 eggs in all, will take her perhaps as long as a month.

Even then, her work is not done. Over the next period of perhaps more than half a year (depending on the water temp-erature), she must make sure the eggs survive until the offspring emerge. She constantly waves her arms gently over the plaits of eggs, making sure that nothing harmful settles on them. With her siphon, she blows water gently over them to keep them aerated. She has probably already built up a defensive wall of rocks outside her den, so that it's difficult for humans to see what's going on inside, but she also uses her arms to keep potential predators away from the eggs, and as far away from the den as possible. This is difficult, though, since she normally does not leave the den at any time.

Throughout this whole period of more than half a year, she never eats. Some scientists believe that her optic glands, behind her optic lobes, have secreted molecules that keep her from feeding. All of the energy in her body is slowly consumed by her work until, by the time the offspring emerge, she has nearly starved to death.

Some divers have experimented with this behavior by bringing food to the female octopus while she is protecting her eggs. She will not eat. Even females accustomed to receiving food from human hands will refuse the food. Researchers speculate that this starvation occurs because food in the den could lure other predators, or because the debris from eating could bring parasites or other kinds of infection that might harm the eggs.

At the end of brooding, when the offspring emerge from their eggs, the mother urges them out of the den and on their way out into the open sea by gently blowing water over them with her siphon. It is likely that only one or two or three of all those carefully nurtured tens of thousands of eggs will survive to adulthood and to reproductive age. Nevertheless, the mother keeps gently siphoning them off into their future in the wide-open sea.

Then she dies, having starved herself to death.

ONE LUCKY SUCKER

Nerve cells firing is what gives us consciousness.

—VINCENT PIERIBONE, YALE UNIVERSITY NEUROBIOLOGIST

T he first giant Pacific octopus I personally encountered was Greg, a young female weighing a mere eight pounds. Greg was not a bit shy. As soon as I mounted the few steps up to her tank at the Aquarium of the Pacific, she came right over and began exploring my arm. Her arms were small, although they didn't seem so to me at the time, since I had nothing to compare them to. At first, she explored my arm, then drew back down into the water, perhaps to process whatever information she had gained.

I turned to talk to James Wood, a marine biologist and passionate sea life enthusiast who devotes his time to the job of chief educator at Southern California's Aquarium of the Pacific, and who was at that moment telling me that we know surprisingly little about the natural history of the giant Pacific octopus.

"The world is filled with big, obvious, huge things that are still mysterious," he said, also mentioning the giant squid and the colossal squid. Wood is an octopus man, and I thought I detected from him a small note of jealousy at all the media attention gained by the dangerously glamorous squid.

The Aquarium of the Pacific, he told me, once had a small two-spot octopus named Lucky Sucker. This octopus was found several miles away from the ocean, walking along a sidewalk in Long Beach, California. The "Lucky" in Lucky Sucker is due to the fact that the right person found her. Lucky was scooped up with a notebook by a concerned student, who then boarded a local bus and carried the octopus all the way to the aquarium, where she was eagerly greeted by staff who knew just how to care for her.

I asked how Lucky could have journeyed so far from the ocean.

"Not sure," Wood answered. "Maybe someone caught it and it escaped." Maybe she was dropped out of a refrigeration truck. Maybe she was bycatch.

Lucky Sucker became one of the aquarium's star performers and, having died years ago, now holds an exalted place in the institution's mythology. Most octopuses are notoriously shy, but Lucky was the Greta Garbo, the great and inscrutable star, of the cephalopod world. If there were an *InStyle* for invertebrates, Lucky would certainly have had her picture on the cover many times. Whenever the education staff had a group of children visiting, they knew they could depend on Lucky to come out and strut her stuff. She would frequently walk around in front of the children like an actress in her screen debut.

I said I was sorry not to have met Lucky.

Wood said that most humans will never meet any octopus at all, even if they spend a lot of time at sea. They're just plain hard to see. When he was a kid, he used to go octopus hunting at 3 a.m., the most likely time for an octopus to be out and about.

Once while he was hunting in the Florida Keys, a lobsterman thought Wood was stealing from his traps. What other reason would the kid have to be out diving at that hour?

When the guy yelled at him, Wood answered that he wasn't after lobster—he was after octopus.

"There's no octopus here," the fisherman said.

"I've seen twenty-one in the past hour," Wood answered. You gotta know where to look.

· : · : · · : · : ·

As we talked, Greg (so named because aquarium staff at first mistook her for a male) decided to renew her acquaintance with my arm.

This time, I visibly flinched as my flesh was squeezed by some of the larger suckers.

"Squeamish?" Wood said.

He seemed genuinely surprised.

Few people in the world are as passionate about cephalopods as Wood. Now in his mid-thirties, he grew up near the Florida coastline and remembers his childhood as that of a "geeky surfer." For as long as he can remember, he caught things out of Florida's polluted canals and brought them home to keep in his bedroom. His parents indulged him, but also found this a not-overly-attractive hobby. Octopuses in particular do not like to stay in tanks, in full view of people. One octopus disappeared almost as soon as Wood brought it home. Wood spent several days looking for it all over the house, worried about what his father would say when he learned that the animal was missing. Then Wood found the animal, still in the tank but hidden away in a tiny crevice where Wood hadn't thought to look.

"How the heck did it do that?" he thought to himself. This would be the first of many such enlightenments. With a child-hood spent on the water and a continuing fascination with the octopus, Wood's route to becoming a cephalopod scientist was not at all convoluted. It came as no surprise that he specialized in marine biology, or that he wrote his doctorate on the mysteries of deep-sea octopuses, or that his career focuses on teaching people about the marvels of the ocean. Wood is among a growing number of experts, in a surprisingly wide range of fields, who believe that the octopus, with its widely distributed net of nerves, may possess a level of intelligence on a par with that of some mammals. If that statement sounds hesitant and full of qualifiers—it is. Most scientists who think about this question say they believe an animal like the giant Pacific octopus is intelligent, but then almost always add: "Of course, we don't really know."

That's their point. Even the giant Pacific octopus, known to so many aquarium visitors around the world, is a mystery. Without knowing more about these animals in their natural habitat, we'll probably not be able to truly evaluate their abilities. "Only those scientists who try to learn everything there is to know about a particular animal have any chance of unlocking its secrets," writes primatologist Franz de Waal, whose lifelong study of chimps and bonobos has revealed remarkable facts about primate culture. Even more challenging would be discovering an appropriate method for measuring cephalopod intelligence. We don't know much about how to evaluate intelligence in any other animal, in fact. Some researchers have begun to write that our failure to recognize "intelligence" in other animals is more a failure of our *own* intelligence than a failure of theirs.

If he were alive today, essayist and octopus observer Gilbert Klingel would probably welcome this new point of view. "Like man, the modern cephalopods have been thrown upon the world naked and without the armor protection of their ancestors," he wrote. In other words, cephalopods, like humans, therefore have to rely on braininess for defense.

James Wood agrees. He said: "We associate intelligence with mammals, with animals that are like us and that are longer-lived. I don't think evolution really cares whether you're a mammal or not. If you have an advantage that helps you survive into the next generation, then that's enough. We're just very human-centric, and believe that what we have is better than anything else."

Wood imagined creating an IQ test—*for* humans, *by* an octopus: "So, the octopus thinks, 'All right, I'm going to make an intelligence test for humans, because they show a little bit of promise in a very few ways.' And the first question the octopus comes up with is this: 'How many color patterns can your severed arm produce in one second?'"

· : · · : · :·

So whose rules do we play by?

One of the reigning theories regarding intelligence is that the quality evolved as a response to social living. Briefly put, primates are smart because they have to learn how to get along with one another. The theory holds that they have to be socially intelligent to wield power over one another in order to get the best or the most food and other items of interest like sex.

Dutch primatologist Carel van Schaik calls this theory "Machiavellian," and postulates instead that intelligence is derived from "social learning." He and many primatologists speculate that intelligence has cultural roots. Van Schaik made an international reputation when he discovered that orangutans can behave socially, something that had not been realized, since the animals usually live independently of one another. He also showed that primates were able to adapt their behavior to circumstances. He attributes their intelligence and ability to exercise social skills to the fact that infant orangutans spend thousands of hours learning from their mothers.

But if researchers come to believe that the giant Pacific octopus is an intelligent being, then the theory that intelligence depends on social interaction will be less viable. The giant Pacific octopus is a solitary animal, so much so that after the mother oversees the hatching of her offspring, she dies. There is no ongoing teaching. Supposedly, everything the newly hatched octopus needs to survive is hardwired into its brain. Yet the animal clearly learns. Throughout its short life, the octopus continuously improves its ability to solve novel and challenging problems like opening jars. These are problems that it might never encounter naturally in the ocean. As we continue to learn about cephalopods, it's likely that our understanding of what it means to be intelligent will expand.

However, judging an animal's "intelligence" by its ability to learn is a dangerous thing, because an animal's ability to learn will depend greatly on what, exactly, is the relevance of what it's learning. Most animals can learn fairly easily what's edible and what isn't, but many species will have trouble with math. We humans cannot change the color or texture of the skin on our arms, but that's very relevant if you're an octopus—so, most likely, we would appear dazed and doltish to an octopus or a squid.

I asked Bill Gilly if he agreed with Scott Brady's joke that squid are the jocks of the cephalopod world, while the octopuses are the intellectuals.

Gilly returned the joke.

"No, I would not completely agree. . . . There are many wimpy, girlie-man squid that are flabby and not what we ordinarily think of as highly athletic," he responded.

He brought up the possibility that we vertebrates might simply be self-centered, or possibly even rather conceited. Gilly said: "I think it is we who are not intelligent enough to devise an IQ test for cephalopods that does not associate humanoid-like activity with intelligence."

Then he stood up for his own research subjects, who, in his opinion, get a bum rap in the octopus-versus-squid discussion. Gilly maintains that humans in general prefer octopuses over squid because we see ourselves as having more in common with the lifestyle choices made by octopuses: "An octopus lives on a two-dimensional substrate, can travel a well-practiced route to go to work on its night-shift job, typically has a house, regularly takes out its trash, and loves to eat crab. So we think the octopus is intelligent because it behaves like we do."

In other words, ho-hum. Not such a challenging lifestyle. We can relate to the octopus's calm, middle-class existence better than to a squid's mysterious, gypsylike wanderings through the ocean depths. The octopus seems to be a thoroughly modern being, willing to trade excitement for security. From the view-point of animals (and people) who live adventurous lives, the octopus opts for boredom.

Said Gilly: "A squid, on the other hand, especially one like *Dosidicus*, lives in a three-dimensional world with boundaries set by temperature, light, oxygen, and salinity rather than physical objects. They do not have permanent places of residence and are nomadic hunters. They eat mesopelagic [mid-level oceanic] organisms that most people don't even know about.

"In short, they are a life-form quite alien to us, and so I think we tend to think of them as being less advanced or intelligent. Again I think that attitude reflects our limitations of perception and understanding. This is just the anthropomorphic nature of man."

Take that, you octopus fans!

· : · : · : · :

Then Gilly concluded: "My less spiritual answer would be that both the squid and the octopus have very large brains. . . .

We don't know much at all about how the cephalopod brain works—there simply have not been many people studying cephalopod brain structure and function in comparison to the vast number studying vertebrates over the past hundred years. Maybe someday we'll learn enough to answer your simple question from a better platform of knowledge."

Octopus man James Wood agrees. "I used to think that the answer to the question was octopus. But now I think that octopus are just easier to manage. Squid are difficult," he said. Most species of squid (except for Margaret McFall-Ngai's "couch potato" Hawaiian bobtail squid) cannot be kept alive for long in captivity. The squid's giant axon, which has helped us understand our own brains, evolved as a fight-or-flight mechanism that allowed them to quickly scoot away from potential harm. Whenever something unexpected happens—someone walks near the squid tank, for example—that giant axon sends a message to the muscles involved in swimming, and the animal immediately darts away. In a tank, this tendency to dart means that the animal may quite often slam into the side of the tank and harm itself.

Squid can learn to overcome this reaction somewhat, as shown by several scientists like the Georgia Aquarium's chief science officer, Bruce Carlson (who trained some squid to feed from his hand when he was in Hawaii), but no one has ever been able to habituate a squid to the degree that Wilson Menashi has habituated the New England Aquarium's giant Pacific octopus, Truman. It's unlikely, given the squid's biology, that a squid will ever playfully interact with a human.

What does that mean about the squid's intelligence? I asked Wood's opinion.

"Let's say you have two people and one of them is really good at math, and the other is an amazing artist," he answered. "Which one is more intelligent? And does your answer depend on whether you're an art teacher or a math teacher?"

· : · : · : ·

This more flexible view of animal intelligence is an emergent phenomenon. The old view was hierarchical with human beings (surprise, surprise) at the top and with the rest of animal life in descending rank. Today animal researchers look less at "levels" of intelligence—"smarter than"—than at styles of intelligence and expressions of a variety of intelligences.

I spoke about this with neuroethologist Paul Patton on the phone one crisp early October day from his office at Bowling Green State University in Ohio. Patton studies the lateral line sense, the sense of water flow, possessed by fish but lacked by humans. Fish can perceive the world around them by using this sense. Even fish that are blind seem to have this alternative ability to "see" their surroundings in this way. It's possible to imagine that a fish would perceive us as quite dim-witted if its only knowledge of us was when we were in the water swimming alongside it.

"Are fish smarter than us, then?" I asked.

"In the sea, perhaps," Patton answered.

With this change in view, the science of teasing apart aspects of another species' intelligence has also changed. As Patton explained in "One World, Many Minds," an article written for *Scientific American*, "complex brains—and sophisticated cognition [thinking]—have evolved from simpler brains multiple times independently" in groups of animals that are, evolutionarily, quite different from each other. Patton particularly noted cephalopod brains.

Of course, if you consider the similarity between the human and the cephalopod neuron, the idea that intelligence could evolve in many different kinds of animals, social or otherwise, doesn't seem quite so surprising. As with eyes, the first step in developing the basic framework has been there all the time. Previously, we just didn't have enough knowledge ourselves to understand that fact.

That's probably why a respected scientist like Alfred Romer could write in 1955 that the brains of birds, complex though they are, allowed for "little learning capacity." If he were alive today, he'd probably be embarrassed by his statement, given all the recent discoveries about how truly brilliant some birds are—if studied in their own environmental niches. Today researchers like Bernd Heinrich in *Mind of the Raven* have shown that some bird species can think and reason. Ravens, Heinrich claims, have even designed "an elegant system of food sharing." Traditionalists have long claimed that what sets humans apart from other animals is that we have "consciousness." Heinrich defines that quality as "the routine mental representation of things and events not directly before the senses," and he believes that ravens possess this ability.

One thing is certain: The pattern of learning in birds is sometimes similar to our own, but not always. As researchers have pursued this avenue, they have discovered that learning in any species is vastly more complicated than twentieth-century scientists realized. Recently, at his animal behavior lab at the City College of New York, researcher Ofer Tchernichovsky discovered one example of just how complex learning can be. He looked at the intricate process by which male zebra finches learn to sing their songs. He found that learning in the zebra finch is a mysterious and many-layered process.

I visited Tchernichovsky's lab to see the finches.

He showed me one of the most amazing animal videos I've ever seen. It showed a male zebra finch who had been isolated from other finches throughout its life. The young bird had never heard another finch sing.

As I saw in the video, at a certain point in the young bird's life, Tchernichovsky gave him access to a recording of a mature male singing. To hear the recording, the young bird had to use an ingenious kind of "switch," a string that he could pull with his beak. At first the young bird was somewhat agitated by the string, a new item in his cage. But then, using his beak, he yanked on it. When the young bird heard the older male finch's song for the first time, he seemed to go into a state of shock. His entire body seemed to go into a trance, as if he had been hypnotized. The bird immediately sank down on his perch and fell into a short, deep sleep. The song was a powerful lullaby.

Then, when the bird woke after a *full* night's sleep, he began to try to sing the song he had heard. His first attempts weren't that good. But over time, after repeatedly hearing the song, and after repeating and repeating and repeating the notes, and after night after night of sleeping—a short nap wasn't enough—his song became more and more like the song of the older male.

"The bird cannot learn to sing this song without the full night's sleep," Tchernichovsky told me.

Tchernichovsky has uncovered layer upon layer of complexity in the song-learning process. He has found out that male birds living near each other and hearing each other then develop a cultural song. But he has also found that individual birds within the group may develop their own unique version of the cultural song. And he has found that over the course of several generations, an isolated group's song will gradually come to resemble more and more the generic "zebra finch song" shared by all male zebra finches. In other words, the basic template for the correct song must be somehow encoded in the bird's brain, just as many researchers believe the basic template for language is encoded in human brains. But there is also room for culture and for individuality.

· : · : · : ·

Is there such a thing as squid culture? Are cuttlefish capable of learning? Is there room for individuality among octopuses? Are they self-aware or even conscious, whatever that might mean? Certainly, the evolutionary biologist and respected animal behavior researcher Martin Moynihan thought so. Moynihan studied the intelligence and social behavior of primates and of birds, but he was also interested in cephalopods, long before others in his field. He wrote before his death at sixty-eight in 1996 that octopuses, squid, and cuttlefish are capable of taking actions that "are overt and decisive. I cannot believe that they are not deliberate and in some sense conscious." Moynihan had spent countless hours diving and following schools of Caribbean reef squid. He concluded that the behavior of individual animals within the school seemed self-aware and even intentional. But how would you find out?

Most of the researchers I spoke with during the writing of this book do suspect that cephalopods are exceptionally intelligent and that they are certainly the most highly developed of invertebrate species. A few suggest that they may be the most intelligent animals in the sea, save for marine mammals. Several researchers believe that intelligence in cephalopods is only logical, since intelligence is a strategy for successful predation, like the teeth of the saber-toothed tiger or the snapping jaws of a crocodile. An octopus, for example, must hunt for its food, and since the hunt involves active investigation and searching for food under rocks and in nooks and crannies like an Easter egg hunt, the evolution of some type of intelligence shouldn't be surprising, despite its asocial lifestyle. The problem, of course, is that we don't have a clear and commonly accepted definition of "intelligence."

I asked a number of scientists: What is it?

One of my favorite comments came from UCLA neuroscientist David Glanzman. Glanzman studies learning and memory on a cellular level. He uses *Aplysia,* a simple mollusk related only distantly to the complex cephalopods. *Aplysia,* commonly called a "sea hare," has only 20,000 neurons. The animal's simple neural systems make it easier to study the complexities of learning and memory than if he were using cephalopods.

I asked him about learning, memory, and thinking in octopuses.

"I sort of look at the octopus as though it were a Martian," he said. "If we saw a Martian repairing a spaceship, we would say: 'Look, there's an intelligent being!'"

We would do that, he explained, because we would recognize that repairing a spaceship took intelligence. But it's equally possible that the things that cephalopods do require intelligence but we don't recognize it as such.

He said: "Nobody knows how cephalopod intelligence works. And that's what I think would be a really useful scientific enterprise. I come back to the Martian example. If you saw a Martian and you saw intelligent behavior, and you got a glimpse of his brain, my guess is that his brain would not be like ours. So, you would want to know, how does it do the things that we do with a totally different brain? Intellectually, I think this is a fascinating question. I want to know how intelligent these animals are. How do they do it? How do their brains do it?"

Glanzman first became fascinated with octopuses when he visited a scientist at the Beaufort Laboratory in North Carolina. The scientist had split the brain of an octopus and had trained the animal so that when the octopus perceived a white ball on one side of the brain, it grabbed the ball and was rewarded with food. When the octopus was presented with the same ball on the other side of the brain, it received a shock if it grabbed the ball.

"I was stunned by how intelligent that animal was," Glanzman said. When the positively reinforced side of the brain was shown the ball, "one of its arms just shot out and grabbed the white ball. But when it saw the ball with the other side of its brain, it literally cowered."

This work was enough to keep Glanzman interested for the next several decades.

"Does this have a cognitive element?" he wondered as we talked. "Does this animal 'plan' things in its life? By this example alone, you wouldn't know that."

Glanzman believes that by studying questions like these, we would learn important things about our own intelligence and about the evolution of intelligence. "Look: Here's something that has cognition and that seems to be similar to us in that, and it has a brain that's structured completely differently from ours. If that's true—does that enhance the possibility that there might have been cognitive beings that arose extraterrestrially? And maybe that means that if you have life, eventually, if you give it enough time, you would have cognition," he said.

"Do I believe that parallel evolution of intelligence is possible? Absolutely! Absolutely!"

Glanzman's enthusiasm for considering the possibility of cephalopod intelligence took me by surprise, but in fact, scientific excitement over the possibility seemed to be nearly universal. Yet despite this excitement, very few researchers are devoting careers to studying the question. In fact, it's considered a scientific backwater. Because of the difficulty of formulating well-thought-out research protocols, very little funding has been made available.

. : . : . : . : . :

It's not easy to put yourself in the mind of an animal with a different brain, even an animal as familiar to us as the dog. One researcher who seems to have somewhat accomplished that goal is Marc Bekoff, retired from the University of Colorado at Boulder. Whereas James Wood imagined a test for human intelligence designed by an octopus, Bekoff asked himself: What if a dog designed a test for self-recognition? For much of the twentieth century, it was assumed that primates were "higher" animals because they could recognize themselves as unique individuals, separate from the rest of the world. Bekoff decided to find out if a dog could also recognize itself as a unique individual.

Prior to Bekoff's experiment, self-recognition studies had primarily involved having an animal look into a mirror. If the animal recognized its own image, it was said to have "self-recognition," a quality that was considered essential for higher-level thinking. Scientists put a dot on the forehead of a primate study subject, then placed a mirror in front of the animal. Seeing the dot in its reflection, the primate usually touched the dot on its own forehead. Other species tested did not do this. Therefore, reasoned scientists, primates recognize the image in the mirror as being their own, and thus possess a sense of self.

Dogs do not do this. Therefore, scientists reasoned, dogs do not understand themselves as unique individuals. Bekoff disagreed with this conclusion. He reasoned that dogs have fewer neurons involved with vision than do we. And they can't see in color. Therefore, they don't care about vision in the same way that primates care about vision. On the other hand, smell is more important to dogs than it is to primates. Smell, Bekoff reasoned, is something that dogs *do* care about.

Bekoff decided to test his own dog, Jethro. Jethro certainly *behaved* as though he understood himself to be a unique individual. He knew what he wanted. He knew what would cause him pain. He knew what he liked to eat. He seemed to have a strong sense of self.

Bekoff knew, as would anyone who'd ever spent time around a dog, that dogs may not care about mirrors. But they do care about smells. A lot. Try taking a dog for a walk where other dogs have recently roamed. You won't get far. This is not surprising. Whereas we have only about 5 million receptors in our noses connected to neuron bodies in our brains to help us differentiate smells, a dog has more than 200 million.

Bekoff designed what became known as the "yellow snow" experiment. Over a period of years, Bekoff picked up samples of snow that had been urinated on by any number of dogs, including Jethro himself. (Not to worry: He used gloves.) These snow samples were moved to new locations without Jethro watching. Then Bekoff measured the amount of time Jethro spent checking out the various samples. It turned out that Jethro spent almost no time over the sample of his own urine, but considerable time over the other samples. Doesn't this, Bekoff asks, show that Jethro recognized himself?

· : · : · : ·

If it takes creative thinking to design a test like that for a dog, an animal with which we are pretty familiar, designing tests that look at intelligence and self-recognition in cephalopods will take a mammoth effort.

SMART SKIN

*Science is not meant to cure us of mystery, but to reinvent
and reinvigorate it.*

—ROBERT SAPOLSKY, NEUROSCIENTIST

S o it's difficult to say exactly what intelligence is. Scientists, philosophers, and educators have been debating the question, sometimes expansively and sometimes explosively, for several thousand years. Interested parties have yet to achieve a peaceful entente.

That's because, like some of science's most basic but non-quantifiable inquiries, there's a political component attached to the problem. Teasing out aspects of human intelligence that are inherited from aspects that may be socially dependent has a great deal to do with the basic human enigma of why we are alive. Do we have a purpose on earth? Is there a God, and does that entity endow each human being with specific, predetermined intellectual abilities? Is it possible to quantify those abilities scientifically? Wouldn't such quantification lead to a nondemocratic society?

In these matters, reason rarely prevails. Consider, for example, the British Sir Francis Galton, who claimed to have found a way to measure human intelligence and ended up instead giving birth to the horrific pseudoscience of eugenics. Galton claimed that it would be possible to breed for human intelligence; his ideas ended up as part of the foundation for Nazism.

Only a few decades ago, there was a famous debate between two intellectual giants from Harvard—the left-wing evolutionist Stephen J. Gould of Harvard and sociobiologist E. O. Wilson. Very simply put, Wilson believes there is an important heritable aspect to intelligence, while Gould, who died of cancer in 2002, insisted on the importance of social circumstances. Their bitter battle reinvigorated the old nature vs. nurture question (the phrase was coined by Galton, in fact), albeit dressed up in new clothing. In *The Mismeasure of Man,* Gould contended that attempts to quantify human intelligence would lead to rigid social stratification. Many scientists felt quite strongly about this: At one point, E. O. Wilson had a pitcher of water dumped on his head at a scientific conference by a member of the International Committee Against Racism.

There is nothing new under the sun. The Greeks debated pretty much the same issue, couched in different language: Are we subject to the whims of the gods, or are we able to determine ourselves what will happen to us in life? Can we be all that we can be? The prevailing American political and social point of view is that we *are* able to determine our own futures and that intelligence can be developed by hard work and a good education. Given these emotionally charged beliefs and the accompanying politics, the scientific study of human intelligence is sometimes dangerous territory.

So it's not surprising that much of the two-decade-old revolution in our understanding of intelligence has come instead from the less-risky field of animal behavior, where we needn't have philosophies like Marxism and Calvinism at odds with each other. Several very influential academic books have been published in the field, among them a breakthrough work on the subject written decades ago by my friend Don Griffin: *Animal Minds.* Recently, bestsellers like Irene Pepperberg's *Alex and Me,* a book in tribute to Pepperberg's verbally communicative parrot, have popularized some of these ideas and brought the public in on some of the behind-the-scenes debate. Pepperberg believes that her research proved that Alex was more than a talking parrot: He was a *thinking* parrot.

I'm willing to grant that Alex could probably "think" on some level. After all, Mark Norman's coconut-shell-carrying octopus can apparently plan for the future by toting around an emergency shelter, just as we would carry a tent on our backs for a camping trip. But the problem is that just as we can't define "intelligence," we can't define precisely the meaning of "thinking." We just haven't made much progress. There does, though, seem to be a kind of general agreement on the attributes of intelligence and thought, just as there was agreement in the nineteenth century on what electricity could do long before scientists figured out that loose electrons created the phenomenon.

Attributes of intelligence seem to include the ability to learn from experience, to adapt behavior, to solve problems, to plan, and to carry out complex tasks. Intelligence seems to involve the quality of curiosity or willingness to explore—all of these, many researchers agree, appear to be signs of this ephemeral and elusive quality of the mind.

· ː · ː · ː · ·

A meditative cuttlefish

If those are the criteria, then cuttlefish seem to belong on the list of intelligent species. I find cuttlefish as intellectually intriguing as orangutans and chimpanzees, and I can watch them for hours on end. I suspect that I like watching these animals because they watch back. And as they watch, they seem to me to be contemplating. Stand at a cuttlefish window at an aquarium and watch the hovering animals. Unless there's something more interesting going on in their tank—mating, feeding, or a dominance battle—chances are high that these odd little animals will swim up to the glass and notice you noticing them. Like cats, cuttlefish seem to pass the time with eyes half-open, staring out at the world in a kind of Zen meditation trance.

I'm not alone in my fascination. As I traveled to various aquariums to research this book, I noticed that many visitors asked first where the octopus tank was located, then quickly tired of watching the octopus sleep and moved on to the cuttlefish exhibit. Few were familiar with cuttlefish—the first remark was often "What are these?"—but then people often stood hypnotized. After all, we humans love making eye contact not just with each other, but with other species as well.

It's difficult to come away from that two-way visual encounter without the impression that cuttlefish not only *watch* you, but *think about* what they're watching. This may only be an illusion created by our own brain design, but the aquarists who care for the animals feel that sense of "thinking" more and more as they continue to interact with their cuttlefish charges.

At the Georgia Aquarium, the bottom of the cuttlefish exhibit is covered with sandy-colored pebblelike material. It has black-and-white checkerboard patterns in several places. When a cuttlefish hovered just above the pebbles, its skin took on the color and texture of sand and pebbles. If it moved a bit and hovered over one of the checkerboard patterns, squares of black and white appeared on its skin. The neon-light-like skin-color changes took only a second or so. Visitors stood and watched, entranced. When the docent was present to explain the details of the cuttlefish light show, the eager crowd asked question after question.

Fascinated by the expressive W-shaped pupils of the cuttlefish, one visitor asked the docent about what the animals were able to see.

"Their vision is in some ways superior to ours, because they don't have a blind spot," he answered. "But they're color-blind. They can do all these crazy camouflage things, and they can't actually see color."

The most common lay question: "Do they do this on purpose?" Is the color change intentional, an expression of intelligence, a conscious decision? Or is it a skill that is mostly hardwired and represents little more than automatic responses to the setting around them? Scientists around the world are asking themselves the same question, but they are still struggling to find ways to tease out the answer in a scientifically valid way.

I asked Amy Rollinson, the Georgia Aquarium's keeper of cuttlefish and someone who has intimate knowledge of her charges' daily lives, what she thought. "There's a lot going on in those little cephalopod brains," she said. "When I give them a new enrichment, a lot of times there's only one or two that will take the dare. But the others watch, to see what happens. They really are very, very smart animals."

In all my travels to research this book, I didn't find one person familiar with cuttlefish who disagreed with Amy's view, despite the current lack of scientific evidence.

· : · : · : ·

There are more than one hundred species of cuttlefish in the shallower regions of the world's oceans, although none occur naturally on the East Coast of the United States. Cuttlefish are a common food source in some parts of the world. They are generally smaller than octopuses and squid, although some species may be as long as three feet. The cuttlefish body plan is similar to that of squid: They have a large mantle containing most of the necessary body organs like the stomach, eight arms, and two feeding tentacles attached to the head around the mouth area. When cuttlefish hover, watching and seeming to meditate, their comparatively short arms sometimes dangle. They look to me like bearded old sages, contemplating the meaning of the universe. And in fact, several popular movies have created humanoid characters with cuttlefish-like curling-flesh arms in place of beards.

Amy dropped some food into a tank so I could see the cuttlefish eat. Their feeding tentacles flashed out from their hiding place among the dangling arms. With laser-quick energy, the cuttlefish grabbed the morsels of food and brought them back to their mouths. The speed of the motion was hypnotizing. If the animal were human-size, it also might have been frightening. The precision of the aim left no doubt that this animal is closely related to predatory squid. Seeing the cuttlefish, small though they are, grab the food reminded me of the description by Japanese scientist Kubodera of the behavior of the feeding tentacles of *Architeuthis*: that they seemed to coil like a python.

Like their cousins the octopus and the squid, cuttlefish live very short lives before undertaking their mating rituals. Then their flesh begins to decay and fall off, and they die a natural death, unless in their senescent state they're eaten first by marauding whales or dolphins. Despite their short life span, cuttlefish have highly developed capacities for communication, particularly expressed through their skin.

Research from a number of scientists implies that cuttlefish sometimes use their skin the way that we sometimes use our mouths. Because we have a facile tongue, teeth, larynx, and lips, we can form words. Over the eons, we have learned to form those words into sentences, and those sentences into concepts. We communicate these concepts to one another, and learn from the information we get back. Now it turns out that cuttlefish may do the same thing, using their skin instead of tongue, lips, and larynx.

Unfortunately for us, we are able to understand only a glimmer of their language and even the meaning of those glimmers we cannot be entirely sure of. But the little we know easily leads to flights of fancy. As I began to think about the changes in skin color and skin texture as possibly being highly sophisticated language, I imagined some cuttlefish frequenting waters often visited by people. I imagined a group of them contemplating human behavior, and sensing the sound of human speech and wondering what all the noise was about.

Then, suddenly, a few thoughtful cuttlefish have an intellectual breakthrough: The noise made by human mouths is like skin-coding, they realize. "They're communicating!" the startled cuttlefish flash to each other. "Maybe humans are smart!"

· : · : · : · : ·

Jean Boal of Pennsylvania's Millersville University is one of the few daring scientists trying to tackle the daunting question of learning and cephalopod intelligence. Her results have been tantalizing. Boal started by asking herself if she was smart enough to find out what the shape-shifting cuttlefish knows. Would she be able to design an experiment that would reveal whether cuttlefish are capable of particular kinds of learning?

First she tested for social recognition. Being able to recognize various individuals in one's own species is considered a sign of intelligence, particularly among mammals. Boal showed that cuttlefish do not recognize each other as unique, distinctive individuals. Into each of two separate tanks, Boal put one male and one female. Each pair mated. The male began to guard the female. Then Boal exchanged males. Each male was put in the other tank, so that the male and female in each tank had *not* mated. Even though both males were in tanks with females with which they had not mated, they guarded the females, behaving as though they were protecting their own sperm. Boal concluded that the males did not recognize various cuttlefish females as unique individuals. If they had recognized their mates, they wouldn't have wasted energy guarding the wrong female, she concluded. "But they went right on guarding," she told me. "They were responding to their own physiology."

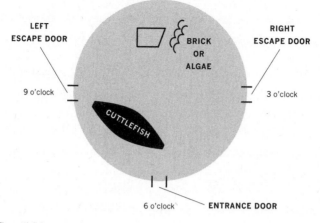

12 o'clock

LEFT
ESCAPE DOOR

RIGHT
ESCAPE DOOR

BRICK
OR
ALGAE

9 o'clock

3 o'clock

CUTTLEFISH

6 o'clock

ENTRANCE DOOR

The cuttlefish maze

This would seem to imply a certain lack of intelligence, but another of Boal's experiments shows that the question is not so easily dismissed. She created a very simple maze containing a problem the cuttlefish had to solve. She released a cuttlefish into a small, round tank with a diameter of only about three cuttlefish body lengths.

If you think of the tank as a clock face, the cuttlefish swam into the tank at the six-o'clock point. To the animal's right, at three o'clock, was one escape door. To the animal's left, at nine o'clock, was another. Each time an animal swam into the tank, one of these doors was open for escape, while the other was closed off with clear plastic. The plastic blocked the escape, but wasn't visible to the cuttlefish.

The first thing an animal saw when it swam into the tank was a "cue" straight ahead at twelve o'clock. The animal had to learn to "read" the cue at twelve o'clock in order to know which way to turn. If the animal swam through the entrance door into the tank and saw algae at the twelve-o'clock point, could it learn to turn right 90 degrees for the escape door? If it saw a brick, could it learn to turn left 90 degrees for the escape door?

"They had to learn to interpret the cue as to which door was going to be open," Boal told me. "And it turned out they could learn to do this. In everyday English, this was an if-then situation: If there's a brick, turn left. If there's algae, turn right."

For human beings, this would be the first step in the development of logic and our ability to use reason in decision-making. I asked Boal what she thought.

"We don't know if they are actually conceiving this the way that we would conceive of it. We would conceive of it as two possibilities: turn right, or turn left. But we don't know if that's what the cuttlefish are doing. You could imagine a robot that's just programmed: Every time I see this, I turn right. Every time I see that, I turn left.

"It would take more working to find out what kind of thinking process the cuttlefish are using, but it does show context sensitivity to their learning. In other words, I don't enter a maze and always turn right. Sometimes I enter a maze and turn right, and sometimes I enter a maze and have to turn left.

"We don't know if they're interpreting the problem using logic. We don't know anything about the thinking process here. All we know is the outcome."

Boal's finding was enticing. Showing that the cuttlefish, easier to study in this way than the squid or octopus, is capable of responding to much more than simple stimulus-response experiences provides a small insight into the cephalopod brain, structured so differently from our own.

While discussing her work, I pointed out to Boal the similarities between our human neurons and cephalopod neurons, and asked if that meant that we might share certain abilities. Or were invertebrates just not capable of doing what we can do with our brains?

"There's a lot of chauvinism about vertebrates and intelligence," she told me. "The concept of intelligence certainly shouldn't be constrained to just vertebrates."

· · ː · ː · ː ·

Perhaps, some suggest, since brains may be products of the evolutionary arms race, intelligent life is a cosmic principle and some characteristics of intelligence, like frustration, may be present in a wider array of animal life than was once expected. This has implications for whatever kind of intelligent life we may find elsewhere in the universe.

But how do we recognize intelligence when we see it in an alien animal? Will we recognize it as such? Or will we have a language barrier, like the communication barrier between humans and cuttlefish? One long-term supposition has been that intelligent life elsewhere will be able to communicate to us through mathematics, supposedly a universal truth.

Perhaps we can practice here on earth by trying to decode some of the basic behaviors of animals like the cuttlefish. Cuttlefish don't do math, at least not a system of mathematics that we understand, although they do use some kind of code with each other. But, suggests comparative psychologist Jesse Purdy of Texas's Southwestern University, maybe we can use some simple responses as a kind of Rosetta stone, a foundation for a deeper understanding of what goes on in the "mind" of a cuttlefish.

Take, for example, frustration. Most of us have experienced that response when we put coins into a vending machine and do not receive a can of soda in return. Many of us consider kicking the soda machine. A few of us actually do it. Most of us eventually just give up and either put more coins in the machine or walk away.

Comparative psychologists have studied frustration for close to a century and have worked out a series of research protocols so that responses from a variety of species can be compared with each other. It turns out that most of the world's species don't seem to experience frustration. Fish, for example, can be trained to strike an object to get a food reward. If the food reward stops coming, the fish will very quickly give up and move on.

But scientists have found that mammals exhibit a strong frustration effect. If their experience tells them that a food reward appears after striking an object, and they don't receive that reward, mammals will continue trying to receive what they expect to receive. The general consensus in the field has been that this frustration drive is mediated by the mammalian limbic system.

But Purdy thinks he may have seen a frustration response in cuttlefish. Purdy began by experiencing his own frustration when he tried to train cuttlefish to swim mazes using the same techniques by which rats are trained. He could not get a consistent response. Every once in a while, he could get one of his subjects to navigate the maze, but never routinely. Without being able to achieve that goal, Purdy was unable to try the long series of research experiments used by scientists to understand some of the basic aspects of an animal's intelligence.

When I asked him if the lack of trainability implied that cuttlefish were not intelligent, his response was quick: "Not at all. Not all animals are set up to solve a maze. A maze isn't something that cuttlefish do for a living. It's going to ambush its food."

But because the cuttlefish don't follow through on behaviors that we know how to study, Purdy was at a loss as to how to proceed. Recently, though, he unearthed a clue that might help. Sometimes a cuttlefish raises its two top arms high above its body and the arms turn a bit red. Purdy believes this behavior is a sign of annoyance or frustration. He had trained a cuttlefish to expect that food would be dropped in its tank when a light in the tank went on and then off. Once when he dropped the food into the tank, it fell behind an object and was out of sight and out of reach of the cuttlefish. The cuttlefish searched for the expected reward, and when it was unable to find it, it raised its two arms above its head.

"The arms turned a deep, dark, blood-red color. If that's a sign of primary frustration, then we might be well on our way to understanding something more about these animals."

The problem, he said, is that we don't know enough about the basic natural history and behavior of cephalopods to be able to formulate the right research questions. Without that understanding, we might be asking the wrong questions, or misinterpreting what we see. Early in his research career, Purdy studied shrimp. His research subjects very quickly learned to strike an object and get a food reward. Within a day or so, the shrimp had learned the task so well that they received at least thirty food rewards in one training session.

But the following day, something strange happened. The shrimp refused to strike the object. No matter what Purdy did, the research subjects ignored the object and did not receive the food reward.

Then he did some reading. It turned out that the shrimp he was training only ate once every thirty days. They didn't strike the object because they were no longer hungry.

"You have to look at things that matter to the animal," Purdy told me. "It all comes back to knowing something about the life history of the animal."

CURIOUS, EXCITING—YET SLIGHTLY DISTURBING

The world was coming of age, and the oceans led the way.

—DORRIK STOW

The cherry blossoms were already in bloom in late February in Portland, Oregon, when Julie arrived to present the final results of her November research cruise in Monterey Bay. It had been an exceptionally short winter on the West Coast. The early burst of color predicted, correctly as it turned out, an unusually warm spring and summer across the continent all the way to the Atlantic. By September, the temperature in Los Angeles would reach 113 degrees.

Julie was speaking at her first major scientific conference, Ocean Sciences 2010. She was anxious but very well prepared. I'd listened to her present at a small conference when I'd met her in Monterey in November. Since then, she'd been coached and drilled by her colleagues until she'd become more confident of both her data and her ability to speak in front of a crowd.

At the Portland conference she spoke to an overflow audience. Scientists crowded around the doorway and in the hall beyond. The work of a graduate student rarely engenders this kind of excitement, but Humboldt behavior was hot science. The audience hoped Julie would provide another clue to the mystery of the squid's sudden proliferation.

She didn't disappoint. She told her listeners about the tagged Humboldt that she had cradled in her arms that November evening and that had turned up seventeen days later west of Ensenada, Mexico. All of the animals she and her team had tagged over several years had been headed in a southerly direction, but that particular animal best demonstrated the speed and perseverance with which these powerful squid were able to travel when so inclined.

Julie also provided her audience with other less obvious but equally important facts. Humboldts have been consistently present in Monterey Bay since 1997, she explained, although their population levels have fluctuated from season to season and from year to year. Small numbers of the animals had been seen in the bay from time to time over many decades, but their presence in fairly large numbers seemed to be something new. Of course, no one knows for sure since accurate fishery records don't exist prior to the twentieth century.

During 2008 and 2009, Julie and the Gilly team had tagged nine of these animals. Julie told her listeners that the team had confirmed, as expected, that Humboldts, like so many other sea species, make a daily migration up to the surface at sunset and back down at dawn. She said that the team found evidence of mated Humboldts during the November cruise but that there was no evidence that the mating had occurred in Monterey Bay. Nor had anyone found newly hatched Humboldts in U.S. waters. The team suspected, but could not prove, that the mating may have occurred in warmer southerly waters. Only one cluster of spawned Humboldt eggs had ever been found—and that was in a warm region near the equator.

Julie also confirmed that the voracious squid were eating fish species that Pacific Coast fishermen catch commercially. Her stomach dissections at John Field's lab had determined that the Humboldts eat hake, rockfish, and smaller squid, sometimes in large quantities.

Julie eventually produced two impressive scientific posters that summed up the research to date. Posters are an important part of the scientific process. During poster sessions at con-ferences, scientists stand beside their posters and wait for others to walk by and stop to read and discuss the information presented. Presenting information in public talks is very important, but in poster sessions scientists can defend their ideas in one-on-one conversations with other experts. These conversations often yield important connections among the work of various labs that might otherwise have gone unnoticed.

In her Portland presentation, Julie developed a well-grounded theory about the Humboldts' arrival in Monterey Bay: Large-scale changes in the earth's ecosystems, including salt water ecosystems, are changing the behavior of the squid.

As the temperature of the planet rises, the chemistry of the ocean is shifting. That shifting chemistry means that life itself will change. We've seen this happen again and again over the planet's four billion years of evolution. No one knows what effect the coming changes will have on life in the ocean. But what is certain is that ocean life *will* change in response to habitat changes. Edinburgh oceanographer Dorrik Stow believes that cephalopods may have lost their outer protective shells and become more mobile in a direct response to one such change in ocean chemistry.

The patterns of ocean currents are also shifting as the oceans warm. This has happened many times over the past four billion years and will likely continue to happen as long as our planet's surface is mostly water. It's inevitable that with those shifts, some sea species will disappear and that other species, perhaps including the predatory Humboldt, will thrive.

Julie predicts that as the oceans warm and land temperatures change surface wind patterns, Humboldts will, at least for a while, become increasingly common along the western coast of North America. Of course, only time will prove her correct. Over the coming years, as she earns her doctorate and begins running a lab of her own, she and a host of other young scientists will continue to gather data and monitor the ocean's ongoing changes.

· : · : · : · :·

Life began in the ocean, perhaps as long ago as four billion years. Only recently did complex life forms colonize the planet's continents. Mollusks may have appeared more than 550 million years ago. Cephalopods definitely appeared by the end of the Cambrian Explosion. Scientists recently recovered a 150-million-year-old squid fossil with an ink sac intact.

We humans appeared only 200,000 years ago, at the tail end of this remarkable saga. On this geologic time scale, other species have appeared and disappeared, sometimes in the blink of an eye. How long we *homo sapiens* will reign is anyone's guess.

There's no guarantee that we'll have the kind of longevity enjoyed by some cephalopod species. I find this slightly disturbing, but at the same time, in an eerie sort of way, rather soothing. Species come and go, but the basic patterns of existence continue.

There's something bigger than ourselves, something barely fathomable to us, given the limitations of our peculiar brains. What is it? Cephalopods, with their vastly different brains and strange neural wiring, may help us find answers to the enigma of our own existence. I find this both curious and exciting.

I began writing this book because I found it marvelous that the same neuron that makes it possible for me to read and write also exists in animals as weird as the octopus, the cuttlefish, and the squid.

As I learned in my research, over the past hundred years we have discovered much about our own minds by studying the brains of cephalopods. Some people find this simple fact of life frightening, or even repugnant. I understand their feelings to a point. That we share so much of our own basic biology with seemingly alien life-forms is a pretty big truth—disconcerting and possibly too large for us to firmly grasp.

But I like the idea. The code of DNA that created the eye that allows me to see has existed, with many variations, for hundreds of millions of years and has given countless species the ability to perceive the world around them. In fact, that very same code allows some cephalopod species to see much more clearly than I do, and it allows some bird and fish species to see many more colors than we humans do. Thus, in some ways, these animals enjoy perceptive abilities that are far greater than our own. I wish I could, for just a moment, enjoy the many colors that the common pigeon or the goldfish sees.

There are some ultimate truths, like the glorious colors that exist in our universe, that my human mind simply cannot grasp.

"What is it like to be a bat?" asked philosopher Thomas Nagel in a famous essay written years after my friend Don Griffin established that most bat species experience the world primarily via sound waves rather than light waves. Nagel concluded that we humans will never be able to fully grasp the mind of a bat. The bat knows things about our planet that we can never know.

It's the same, I suspect, with the octopus, cuttlefish, and squid. But science is managing to provide us with titillating glimpses. The whole picture will always be denied to us because we are, after all, only a tiny part of the ultimate creation (whatever that might be). Curious scientists like Julie and Gilly and the rest of the crew will help us discover more and more of the pieces to the puzzle.

ACKNOWLEDGMENTS

Kraken relied very much on the kindness of strangers, on people willing to give their time to help create a book in which they themselves had no particular stake.

The book is in part a product of the Ocean Foundation, a D.C.-based community foundation that supports the health of the world's oceans. The Ocean Foundation helped provide $10,000 in travel money without which the West Coast portion of this book would not have been possible. Many thanks to Mark Spalding for helping this happen, and to the always supportive and very thoughtful Diane Davidson.

I thank also the patient staff of the Cotuit Library in Cotuit, Massachusetts, who were, as usual, gracious and pleasant about helping me find the sometimes quite strange books I wished to look at.

And I thank Barbara Legg for her helpful suggestions in shaping the book, most particularly for her patience in sticking with this project from beginning to end and for being willing to read endless versions of the manuscript.

And there's of course my husband, Greg Auger, to thank, both for providing many of the photographs in the book and for providing a great deal of logistics support. My cousins (and close friends) Susan Williams and Shirley Smith were also uniquely helpful in their own very special ways.

Marine Biological Laboratory neuroscientist Joe DeGiorgis, a man of great forbearance, spent many days helping me flesh out some of the scientific details in the book. Without him this would have been a very different project.

Other scientists also provided a considerable amount of help, including, of course, the West Coast team in the Gilly lab. Julie Stewart, while earning her doctorate, answered all e-mails and phone calls in a prompt and courteous manner. It can't have been easy for her. Danna Staaf, also earning her doctorate, was helpful as well. Lou Zeidberg was extremely supportive.

And then, of course, there's Gilly himself, a man with a wonderful sense of humor, a finely tuned sensitivity to the power of words, and a depth of knowledge as deep as the Monterey Canyon.

I also want to mention John Field of the National Oceanic and Atmospheric Administration, who is not officially part of the Gilly team but who works closely with those scientists and was more than gracious in helping with the book, including providing a visit to his facilities.

Many other scientists also contributed, including the always cheerful James Wood of the Aquarium of the Pacific in Long Beach, California, who kindly introduced me to my first giant Pacific octopus; UCLA neuroscientist David Glanzman; teuthologist Clyde Roper; Bruce Carlson of the Georgia Aquarium; Amy Rollinson, also of the Georgia Aquarium; UCLA scientist Barney Schlinger; Yale University neuroscientist Vincent Pieribone; the brilliant biologist Margaret McFall-Ngai; neurosurgeon Bruce Andersen, who tried valiantly to teach me how to dissect a squid axon; Scott Brady, a scientific marvel and a very contemplative man; Jesse Purdy, who took the time to explain why humans kick vending machines; James Cosgrove, lifelong giant Pacific octopus observer; Nina Strömgren Allen, who spoke to me from her hospital bed; neuroethologist Paul Patton; Eric Hochberg of the Santa Barbara Museum of Natural History; Roger Hanlon; Todd Oakley; Jennifer Mather; Mike Vecchione; Ofer Tchernichovsky, who showed me one of the most amazing animal learning videos I've ever seen; Marc Bekoff, who has thought so much about the meaning of intelligence in the animal world and who was willing to share some of his ideas with me; and most particularly my friend the late Don Griffin, bat echolocation expert and among the most encouraging and kindest of scientists.

Others who deserve mention include the intrepid Tom Mattusch, who introduced me to Humboldt fishing on the *Huli Cat*; fisherman and high school marine biologist Rob Yeomans, who became a good friend; Wilson Menashi, who continues faithfully to volunteer at the New England Aquarium; the Pelagic Shark Research Foundation's Sean Van Sommeran; Steve Atherton of Newburyport; Bill Papoulias of Newburyport; Greg Early, formerly of the New England Aquarium; Jack Pearce, who kindly shared a lunch and many of his books; and so many others.

Abrams editor David Cashion, who understood the lure of squid from the very beginning, deserves credit for both his patience and his unfailing good humor. He's a great guy to work with. Judy Heiblum of Sterling Lord realized from the outset that although squid are indeed very—very—weird, they're also strangely alluring.

And finally, special thanks to William Breisky, the best editor a young reporter could ever have had.

BIBLIOGRAPHY

Allen, R., N. Allen, and J. Travis. "Video-enhanced Contrast Polarization (AVEC-POL) Microscopy: A New Method Capable of Analyzing Microtubule-Related Motility in Reticulopodial Network of *Allogromia laticollaris*." *Cell Motility* 1 (1981): 291–302.

Allen, Robert Day, et al. "Fast Axonal Transport in Squid Giant Axon." *Science,* New Series, 218, no. 4577 (Dec. 10, 1982): 1127–29.

Alves, Christelle, et al. "Orientation in Cuttlefish *Sepia officinalis*: Response versus Place Learning." *Animal Cognition* 10 (2007): 29–36.

Anderson, Roland C. "Smart Octopus?" *The Festivus* 38, no. 1 (2006): 7–9. http://www.thecephalopodpage.org/OctopusSmarts.php

Anderson, Roland C., and James B. Wood. "Enrichment for Giant Pacific Octopuses." *Journal of Applied Animal Welfare Science* 4, no. 2 (2001): 157–68.

Angier, Natalie. "Listening to Bacteria." *Smithsonian,* July/August 2010: 77–82.

Bahn, Paul G. *Journey Through the Ice Age.* Berkeley: University of California Press, 1997.

Bavendam, Fred. "Chameleon of the Reef." *National Geographic,* September 1995: 94–107.

Beebe, William. *Galápagos: World's End.* New York: G. P. Putnam's Sons, 1924.
———. *Half Mile Down.* New York: Harcourt, Brace and Company, 1934.

Bekoff, Marc. *Animal Passions and Beastly Virtues: Reflections on Redecorating Nature.* Philadelphia: Temple University Press, 2006.

Benchley, Peter. *Beast.* New York: Random House, 1991.

Boal, J. G. "Absence of Social Recognition in Laboratory-reared Cuttlefish, *Sepia officinalis*." *Animal Behavior* 52 (1996): 529–37.

Boal, J. G. "Social Recognition: A Top Down View of Cephalopod Behavior." *Vie et Milieu* 56, no. 2 (2006): 69–79.

Brady, Scott T., Raymond J. Lasek, and Robert D. Allen. "Fast Axonal Transport in Extruded Axoplasm." *Science,* New Series 218, no. 4577 (Dec. 10, 1982): 1129–31.

Brandt, A. V. *Fish Catching Methods of the World.* London: Fishing News Books, 1964.

Broad, William J. "Diving Deep for a Living Fossil." *New York Times,* August 25, 2009, D1.

Budelmann, B. U. "Autophagy in *Octopus.*" *South African Journal of Marine Science* 20, no. 1 (December 1998): 101–8.

Byatt, Andrew, Alastair Fothergill, and Martha Holmes. *The Blue Planet: Seas of Life.* New York: DK Publishing, 2001.

Byrne, Ruth A., et al. "Does *Octopus Vulgaris* Have Preferred Arms?" *Journal of Comparative Psychology* 120, no. 3 (2006): 198–204.

Carroll, Sean B. *Endless Forms Most Beautiful: The New Science of Evo Devo.* New York: W. W. Norton & Co., 2005.

———. *The Making of the Fittest: DNA and the Ultimate Forensic Record of Evolution.* New York: W. W. Norton & Co., 2006.

———. *Remarkable Creatures: Epic Adventures in the Search for the Origin of Species.* New York: Houghton Mifflin Harcourt, 2009.

Carson, Rachel. *The Sea Around Us,* rev. ed. New York: Oxford University Press, 1961.

Cherel, Yves, and Keith A. Hobson. "Stable Isotopes, Beaks and Predators: A New Tool to Study the Trophic Ecology of Cephalopods, Including Giant and Colossal Squids." *Proceedings of the Royal Society* B 272 (2005): 1601–7.

Clapp, Pamela. "Trucking Down the Axon." *MBL Science* Winter 1986, http://www.mbl.edu/publications/pub_archive/Loligo/Clapp/Clapp.html

Conley-Early, Andrea. "The Hunt for Giant Squid," *SeaFrontiers,* http://seawifs.gsfc.nasa.gov/OCEAN_PLANET/HTML/squid_sea_frontiers.html

Cosgrove, James A., and Neil McDaniel. *Super Suckers: The Giant Pacific Octopus and Other Cephalopods of the Pacific Coast.* Madeira Park, BC: Harbor Publishing, 2009.

Cousteau, Jacques-Yves. "Fishermen Explore a New World Undersea." *National Geographic,* October 1952, 431–72.

Cousteau, Jacques-Yves, and Philippe Diole. *Octopus and Squid: The Soft Intelligence.* Garden City, NY: Doubleday, 1973.

Cousteau, Jacques-Yves, with Susan Schiefelbein. *The Human, the Orchid, and the Octopus: Exploring and Conserving Our Natural World.* New York: Bloomsbury, 2007.

Coyne, Jerry A. *Why Evolution Is True.* New York: Viking, 2009.

Dalton, Rex. "True Colors." *Nature* 428 (April 8, 2004): 596–97.

Darmaillacq, Anne-Sophie, et al. "Early Familiarization Overrides Innate Prey Preference in Newly Hatched *Sepia officinalis* Cuttlefish." *Animal Behavior* 71 (2006): 511–14.

Darmaillacq, Anne-Sophie, et al. "Effect of Early Feeding Experience on Subsequent Prey Preference by Cuttlefish, *Sepia officinalis.*" http://www3.interscience.wiley.com/journal/109795599

David, R. W., et al. "Diving Behavior of Sperm Whales in Relation to Behavior of a Major Prey Species, the Jumbo Squid, in the Gulf of California, Mexico." *Marine Ecology Progress Series* 333 (March 12, 2007): 291–302.

Deary, Ian J. *Intelligence: A Very Short Introduction.* New York: Oxford University Press, 2001.

Dickel, Ludovic, Jean G. Boal, and Bernd U. Budelmann. "The Effect of Early Experience on Learning and Memory in Cuttlefish." *Developmental*

BIBLIOGRAPHY

Psychobiology 36, no. 2 (Feb. 24, 2000): 101–10.

Durrell, Gerald. *My Family and Other Animals.* London: Penguin Books, 1959.

Ellis, Richard. *The Search for the Giant Squid: The Biology and Mythology of the World's Most Elusive Sea Creature.* New York: Penguin, 1999.

Faulkner, Doug. "The Chambered Nautilus: Exquisite Living Fossil." *National Geographic,* January 1976, 38–41.

Field, John C., Ken Baltz, et al. "Range Expansions and Trophic Interactions of the Jumbo Squid, *Dosidicus gigas,* in the California Current." *CalCOFI Rep.* 48: 2007.

Finger, Stanley. *Minds Behind the Brain: A History of the Pioneers and Their Discoveries.* London: Oxford University Press, 2000.

Finn, Julian, Tom Tregenza, and Mark Norman. "Defensive Tool Use in a Coconut-carrying Octopus." *Current Biology* 19, no. 22 (December 15, 2009). http://www.cell.com/current-biology/abstract/S0960-9822%2809%2901914-9

———. "Preparing the Perfect Cuttlefish Meal." PloS ONE http://www.plosone.org/article/info%3Adoi%2F10.1371%2Fjournal.pone.0004217

Fisher, Arthur. "He Seeks the Giant Squid." *Popular Science,* May 1, 1995. http://seawifs.gsfc.nasa.gov/OCEAN_PLANET/HTML/ps_roper.html

Gabriel, Otto, Andres von Brandt, Klaus Lange, Erdmann Dahm, and Thomas Wendt. *Fish-catching Methods of the World.* New York: Wiley-Blackwell. 2005.

Gayles, Richard. "Cone Snail Venom Effective Remedy for Pain." November 26, 2009. http://www.spacecoastmedicine.com/2009/11/cone-snail-venom-effective-remedy-for-pain.html

Gilbert, Daniel L., William J. Adelman, and John M. Arnold. *Squid as Experimental Animals.* New York City: Springer Publishing Company, 1990.

Gilly, William F., and Unai Markaida. "Perspectives on *Dosidicus gigas* in a Changing World." In *GLOBEC Report* 24, edited by R. J. Olson and J. W. Young: 81–90.

Gilly, William F., et al. "Spawning by Jumbo Squid (*Dosidicus gigas*) in the San Pedro Martir Basin, Gulf of California, Mexico." *Marine Ecology Progress Series* 313: 125–33.

Goldstein, Miriam. "Motion in the Ocean: The Six Secrets of Squid Sex." *Slate,* March 13, 2009. http://www.slate.com/id/2211343/

Gould, Carol Grant. *The Remarkable Life of William Beebe: Explorer and Naturalist.* Washington, D.C.: Island Press, 2004.

Groff, Bethany. *A Brief History of Old Newbury: From Settlement to Separation.* Charleston, SC: The History Press, 2008.

Guerra, A., et al. "Records of Giant Squid in the North-eastern Atlantic, and Two Records of Male *Architeuthis* sp. off the Iberian Peninsula." *Journal of the Marine Biological Association of the United Kingdom* 84 (2004): 427–31.

Gutfreund, Yoram, et al. "Organization of Octopus Arm Movements: A Model System for Studying the Control of Flexible Arms." *The Journal of Neuroscience* 16, no. 22 (November 15, 1996): 7297–307.

Hanlon, Roger T., and John B. Messenger. *Cephalopod Behaviour.* New York:

Cambridge University Press, 1996.

Hanlon, Roger T., et al. "Adaptable Night Camouflage by Cuttlefish." *The American Naturalist* 169, no. 4 (April 2007): 543–51.

Harvey, Moses. "Gigantic Cuttlefishes in Newfoundland." *The Annals and Magazine of Natural History* ser. 4, 13, no. 73 (1874): 67–69.

Heinrich, Bernd. *Mind of the Raven.* New York: HarperCollins, 1999.

Hochner, Binyamin, Tal Shomrat, and Graziano Fiorito. "The Octopus: A Model for a Comparative Analysis of the Evolution of Learning and Memory Mechanisms." *The Biological Bulletin,* http://www.biolbull.org/cgi/content/full/210/3/308

Hochner, Binyamin, et al. "A Learning and Memory Area in the Octopus Brain Manifests a Vertebrate-Like Long Term Potentiation." *Journal of Neurophysiology* 90 (August 13, 2003): 3547–54.

Hodgkin, A. L., and A. F. Huxley. "Action Potentials Recorded from Inside a Nerve Fiber." *Nature* 144 (October 21, 1939): 710–11.

Holmes, Richard. *The Pursuit.* New York: E. P. Dutton, 1975.

Hoving, Hendrick Jan T., et al. "Sperm Storage and Mating in the Deep-sea Squid *Taningia danae.*" *Marine Biology* 157 (2010): 393–400.

Hugo, Victor. *The Toilers of the Sea.* New York: Modern Library, 2002.

Hvorecny, Laureen M., et al. "Octopuses (*Octopus bimaculoides*) and Cuttlefishes (*Sepia pharaonis, S. officinalis*) Can Conditionally Discriminate." *Animal Cognition* 10(4) (2007 Oct.): 449–59.

"It's a Girl: Atlantic Mystery Squid Undergoes Scrutiny." *Science News* 171 (March 17, 2007): 165.

Jackson, G. D., and N. A. Moltschaniwsky. "Analysis of Precision in Statolith Derived Age Estimates of the Tropical Squid *Photololigo* (Cephalopoda: Loliginidae)." *ICES Journal of Marine Science* 56 (1999): 221–27.

Kettmann, Matt. "What's 30 Feet Long with Eight Legs, a Big Beak, and a Life of Mystery? Giant Squid Dissected at the Santa Barbara Museum of Natural History." *The Santa Barbara Independent,* August 22, 2008, http://www.independent.com/news/2008/aug/22/whats-30-feet-long-eight-legs-big-beak-and-life-my/

Klingel, Gilbert C. *Inagua: An Island Sojourn.* New York: Dodd, Mead and Co., 1940.

Klocel, Roger. "The Most Interesting Animal Alive." *Aquaticus: Journal of the Shedd Aquarium* 24, no. 2 (1994).

Knecht, G. Bruce. *Hooked: Pirates, Poaching, and the Perfect Fish.* New York: Rodale Books, 2006.

Koslow, Tony. *The Silent Deep: The Discovery, Ecology, and Conservation of the Deep Sea.* Chicago: University of Chicago Press, 2007.

Kubodera, Tsunemi, Yasuhiro Koyama, and Kyoichi Mori. "Observations of Wild Hunting Behaviour and Bioluminescence of a Large Deep-sea, Eight-armed Squid, *Taningia danae.*" *Proceedings of the Royal Society* B 274 (2007): 1029–34.

Kubodera, Tsunemi, and Kyoichi Mori. "First-ever Observations of a Live Giant Squid in the Wild." *Proceedings of the Royal Society* B 272 (2005): 2582–86.

doi: 10. 1098/rspb.2005.3158.

Kunzig, Robert. *Mapping the Deep: The Extraordinary Story of Ocean Science.* New York: W. W. Norton & Co., 1999.

Lane, Frank W. *Kingdom of the Octopus.* New York: Sheridan House, 1960.

Linden, David J. *The Accidental Mind: How Brain Evolution Has Given Us Love, Memory, Dreams, and God.* Cambridge, MA: The Belknap Press of Harvard University, 2007.

Lindsey, Rebecca. "Ancient Crystals Suggest Earlier Ocean." NASA Earth Observatory Web site. http://earthobservatory.nasa.gov/Features/Zircon/

Livingstone, Margaret. *Vision and Art: The Biology of Seeing.* New York: Harry N. Abrams, 2002.

MacInnis, Joseph. *Aliens of the Deep.* Washington, D.C.: National Geographic Society, 2004.

Madsen, Axel. *Cousteau: An Unauthorized Biography.* New York: Beaufort Book Publishers, 1987.

Marshall, Charles R., and David K. Jacobs. "Flourishing After the End-Permian Mass Extinction." *Science* 325, no. 28 (August 2009): 1079–80.

Mather, Jennifer A. "Cephalopod Consciousness: Behavioural Evidence." *Consciousness and Cognition* 17 (2008): 37–48.

———. *Octopus: The Ocean's Intelligent Invertebrate.* Portland, OR: Timber Press, 2010.

Mathger, Lydia M., Nadav Shashar, and Roger Hanlon. "Do Cephalopods Communicate Using Polarized Light Reflections from Their Skin?" *Journal of Experimental Biology* 212 (2009): 2133–40.

McFall-Ngai, Margaret. "Adaptive Immunity: Care for the Community." *Nature* 445 (January 11, 2007): 153.

———. "Are Biologists in 'Future Shock'?: Symbiosis Integrates Biology Across Domains." *Nature Reviews: Microbiology* 6 (October 2008): 789–92.

———. "Love the One You're With: Vertebrate Guts Shape Their Microbiota." *Cell* 127 (October 20, 2006): 247–49.

Menard, Wilmon. "Octopus Wrestling Is My Hobby." *Modern Mechanix* 161 (April 1949): 76–79.

Miserez, Ali, et al. "Microstructural and Biochemical Characterization of the Nanoporous Sucker Rings from *Dosidicus gigas.*" *Advanced Materials* 20 (2008): 1–6.

Miserez, Ali, et al. "The Transition from Stiff to Compliant Materials in Squid Beaks." *Science* 319, no. 5871 (March 28, 2008): 1816–19.

Monks, Neale, and Philip Palmer. *Ammonites.* London: The Natural History Museum, 2002.

Morfini, Gerardo A., David L. Stenoien, and Scott T. Brady, "Axonal Transport," in *Basic Neurochemistry: Molecular, Cellular, and Medical Aspects,* edited by George Siegel, 7th ed.: 485–502. Amsterdam and Boston: Academic Press, November 14, 2005.

Morfini, Gerardo A., et al. "Axonal Transport Defects in Neurodegenerative Diseases." *The Journal of Neuroscience* 29, no. 41 (October 14, 2009): 12776–86.

Moynihan, Martin H. "Self-awareness, with Specific References to Coleoid Cephalopods." *Anthropomorphism, Anecdotes, and Animals.* Albany, NY: State University of New York Press, 1996.

Nichols, Peter. *A Voyage for Madmen.* New York: HarperCollins, 2001.

Nouvian, Claire. *The Deep: The Extraordinary Creatures of the Abyss.* Chicago: The University of Chicago Press, 2007.

Overbye, Dennis. "A Sultry World Is Found Orbiting a Distant Star." *New York Times,* Dec. 17, 2009.

Owen, James. "'Colossal Squid' Revives Legends of Sea Monsters." *National Geographic News,* April 23, 2003. http://news.nationalgeographic.com/news/pf/54118877.html

Packard, A. S. "Colossal Cuttlefishes." *The American Naturalist* 7, no 2. (February 1873): 87–94.

Palumbi, Stephen R., Karen L. McLeod, and Daniel Grunbaum. "Ecosystems in Action: Lessons from Marine Ecology about Recovery, Resistances, and Reversibility." *BioScience* 58, no. 1 (January 2008): 33–42.

Paquita, E. A., et al. "Saving Darwin's Muse: Evolutionary Genetics for the Recovery of the Floreana Mockingbird." *Biology Letters* (November 18, 2009). http://rsbl.royalsocietypublishing.org/content/early/2009/11/17/rsbl.2009.0778.full.pdf+html

Parker, Andrew. *In the Blink of an Eye.* New York: Perseus Publishing, 2003.

Patton, Paul. "One World, Many Minds: Intelligence in the Animal Kingdom." http://www.scientificamerican.com/article.cfm?id=one-world-many-minds

Pereira, Joao, et al. "First Recorded Specimen of the Giant Squid *Architeuthi* sp. in Portugal." *Journal of the Marine Biological Association of the United Kingdom* 85 (2005): 175–76.

Pieribone, Vincent, and David F. Gruber. *Aglow in the Dark: The Revolutionary Science of Biofluorescence.* Cambridge, MA: The Belknap Press of Harvard University, 2003.

Pinti, Daniele L. "The Origin and Evolution of the Oceans," in *Lectures in Astrobiology* 1, pt. 1. New York: Springer, 2006, 83–112.

Poirier, R., R. Chichery, and L. Dickel. "Effects of Rearing Conditions on Sand Digging Efficiency in Juvenile Cuttlefish." *Behavioral Processes* 67 (2004): 273–79.

Quammen, David. *Natural Acts: A Sidelong View of Science and Nature.* New York: Schocken Books, 1985.

Rehling, Mark J. "Octopus Prey Puzzles." *The Shape of Enrichment* 9, no. 3 (August 2000). http://www.torontozoo.com/meet_Animals/enrichment/Files/Octopus%20Enrichment.pdf

Rincon, Paul. "New Giant Squid Predator Found." *BBC News,* January 8, 2004. http://news.bbc.co.uk/2/hi/science/nature/3370019.stm

Robb, Graham. *Victor Hugo: A Biography.* New York: W. W. Norton & Co., 1997.

Roberts, Callum. *The Unnatural History of the Sea.* Washington, D.C.: Shearwater Books, 2007.

Roeleveld, Martina A. C. "Is There a Squid in Your Future? Perspectives for

New Research." *American Malacological Bulletin* Special edition no. 1 (1985): 93–100.

———. "Tentacle Morphology of the Giant Squid *Architeuthis* from the North Atlantic and Pacific Oceans." *Bulletin of Marine Science* 71, no. 2 (2002): 725–37.

Roper, Clyde. "Tracking the Giant Squid: Mythology and Science Meet Beneath the Sea." *Wings* 21, no. 1 (Spring 1998): 12–17.

Roper, Clyde F. E., and Kenneth J. Boss. "The Giant Squid." *Scientific American* 246, no. 4 (1983): 96–105.

Roper, Clyde F. E., and Richard E. Young. "First Records of Juvenile Giant Squid, *Architeuthis* (Cephalopoda: Oegopsida)." *Proceedings of the Biological Society of Washington* 85, no. 16 (August 30, 1972): 205–22.

Rozwadowski, Helen M. *Fathoming the Ocean: The Discovery and Exploration of the Deep Sea.* Cambridge, MA: The Belknap Press of Harvard University, 2005.

Ruby, Edward, Brian Henderson, and Margaret McFall-Ngai. "We Get by with a Little Help from Our (Little) Friends." *Science* 303 (February 27, 2004): 1305–7.

Ruse, Michael, and Joseph Travis, eds. *Evolution: The First Four Billion Years.* Cambridge, MA: The Belknap Press of Harvard University, 2009.

Sagarin, Raphael D., William F. Gilly, et al. "Remembering the Gulf: Changes to the Marine Communities of the Sea of Cortez since the Steinbeck and Ricketts Expedition of 1940." *Frontiers in Ecology and the Environment* 6, no. 7 (2008): 372–79.

Scales, Helen. *Poseidon's Steed: The Story of Seahorses from Myth to Reality.* New York: Gotham, 2009.

Science Daily. "Snail Toxin May Spur New Drugs for Alzheimer's, Parkinson's, Depression." Aug. 21, 2006. http://www.sciencedaily.com/releases/2006/08/060820191735.htm

Scully, Tony. "The Great Squid Hunt." *Nature* 454 (August 21, 2008): 934–36.

Shillinglaw, Susan. *A Journey into Steinbeck's California.* Berkeley: Roaring Forties Press, 2006.

Shubin, Neil. *Your Inner Fish: A Journey into the 3.5-Billion-Year History of the Human Body.* New York: Vintage, 2009.

Shubin, Neil, Cliff Tabin, and Sean B. Carroll. "Deep Homology and the Origins of Evolutionary Novelty." *Nature* 457 (February 12, 2009): 818–23.

Staaf, Danna, et al. "Natural Egg Mass Deposition by the Humboldt Squid (*Dosidicus gigas*) in the Gulf of California and Characteristics of Hatchlings and Paralarvae." *Journal of the Marine Biological Association of the United Kingdom* 88, no. 4 (2008): 759–70.

Steinbeck, John. *Cannery Row.* New York: Viking, 1945.

———. *The Log from the Sea of Cortez.* New York: Viking, 1951.

Stewart, Julie, and William F. Gilly. "Piscivorous Behavior of a Temperate Cone Snail, *Conus californicus.*" *The Biological Bulletin* 209 (October 2005): 146–53. http://www.biolbull.org/cgi/content/full/209/2/146

Stow, Dorrik. *Vanished Ocean: How Tethys Reshaped the World.* New York:

Oxford University Press, 2009.

Strugnell, Jan, et al. "Neotenous Origins for Pelagic Octopuses." *Current Biology* 14, no. 8 (April 20, 2004): R300–R301.

Treffil, James, Harold J. Morowitz, and Eric Smith. "The Origin of Life." *American Scientist* 97 (May/June 2009): 206–13.

Tucker, Abigail. "King of the Sea." *Smithsonian*, July/August 2010: 27–34.

Van Schaik, Carel. "Why Are Some Animals So Smart?" *Scientific American* (April 2006).

Vecchione, Michael, and Richard Young. "The Squid Family *Magnapinnidae* (Mollusca: Cephalopoda) in the Atlantic Ocean, with a Description of a New Species." *Proceedings of the Biological Society of Washington* 119, no. 3 (2006): 365–72.

Vecchione, Michael, et al. "Worldwide Observations of Remarkable Deep-sea Squid." *Science* 294: 2505–6.

Voss, Gilbert L. "Shy." *National Geographic*, December 1971: 776–99.

———. "Squids: Jet-Powered Torpedoes of the Deep." *National Geographic* (March 1967): 385–411.

Wade, Nicholas. "From One Genome, Many Types of Cells. But How?" *New York Times*, February 22, 2009, D-4.

Walker, Sally M. *Fossil Fish Found Alive: Discovering the Coelacanth.* Minneapolis: Carolrhoda Books, 2002.

Walton, John. *Storied Land: Community and Memory in Monterey.* Berkeley: University of California Press, 2001.

Ward, Peter Douglas. *In Search of Nautilus.* New York: Simon & Schuster, 1988.

Weare, Nancy V. *Plum Island: The Way It Was.* Newbury, MA: Newburyport Press, 1996.

Wells, M. J. *Octopus: Physiology and Behaviour of an Advanced Invertebrate.* London: Chapman and Hall, 1978.

Wilford, John Noble. "From Arctic Soil, Fossils of a Goliath That Ruled the Jurassic Seas." *New York Times*, March 17, 2009, D-3.

Wong, Kate. "Shell Game." *Scientific American*, February 2009: 26.

Young, John Zachary, "Cephalopods and Neuroscience." *The Biological Bulletin* 168 Suppl. (June 1985): 153–58.

Young, Richard Edward, and Clyde F. Roper, "Bioluminescent Countershading in Midwater Animals: Evidence from Living Squid." *Science* 191 (March 12, 1976): 1046–48.

Ziedberg, Louis David. "First Observations of Sneaker Mating in the California Market Squid, *Doryteuthis opalescens* (Cephalopoda: Myopsida)." *Journal of the Marine Biology Association* 2–Biodiversity Records 5998. http://www.mba.ac.uk/jmba/biodiversityrecords.php?jmbaref

Zeidberg, Louis D., and Bruce H. Robison. "Invasive Range Expansion by the Humboldt Squid, *Dosidicus gigas,* in the Eastern North Pacific." *Proceedings of the National Academy of Science* 104, no. 31: 12948–50. http://www.pnas.org/content/104/31/12948.abstract

Zimmer, Carl. *At the Water's Edge: Macroevolution and the Transformation of Life.* New York: Free Press, 1998.

VIDEOS

"Aliens of the Deep." Walt Disney Pictures, 2005.

Beast. Made-for-television movie, 1996.

"Cuttlefish: Kings of Camouflage." Nova, 2007.

"Devils of the Deep: Jumbo Squid." Creatures of the Deep, National Geographic, 2008.

"Eyeball to Eyeball: An Interview with Clyde Roper." First Person Series, Errol Morris, 2000.

It Came from Beneath the Sea. Columbia Pictures, 1955.

"Quest for the Giant Squid." Discovery Channel, 2000.

Pirates of the Caribbean. Walt Disney Pictures, 2006.

"Science Is Fiction: 23 Films by Jean Painlevé." Criterion Collection, re-release 2009.

"Sea Monsters: Search for the Giant Squid." National Geographic Video, 1998.

CREDITS

INDEX

Page numbers in italics refer to illustrations.

ink sacs, 10, 45; in cephalopods, 25

intelligence: attributes of, 186; and communication, 59, 79; defining, 180–82, 185–86; and evolution, 173–74, 177, 179, 181, 184–85; frustration response, 193–95; and learning process, 155–56, 165, 173–74, 177–79, 180–81, 186, 190–92, 194 (*see also* prey puzzles); measuring, 173, 177, 190; nature-versus-nurture, 184–85; and predation, 180; and social behavior, 173–74

ions, 118–19

iridophores, 90–91

Japan, 69, 72, 74, 82–83; *69*

Japanese flying squid, 22

jellyfish, 18–19, 81, 82, 146

Jorgensen, Salvador, 49

Kaikoura Canyon (New Zealand), 68

Kandel, Eric, 98

Kimberella, 18–20, 21; *19*

kinesin molecule, 128, 129–30, 131

Klingel, Gilbert, 153–54, 173

Kraken. *See* giant squid (Kraken) (*Architeuthis*)

Kubodera, Tsunemi, 62, 69, 72, 189; *69, 73*

La Jolla (California), 36, 37

Loeb, Jacques, 108

Loligo opalescens (California market squid), 51, 56, 136–38

Loligo pealei, 99, 110–11, 136–37; neuroscience research on, 99–100, 105–6, 112, 115–20, 125–28; *99, 105*

"Lucky Sucker" (octopus), 170–71

luminosity. *See* bioluminescence

MacInnis, Joseph, 112

males. *See* mating and reproduction

mantle, 23–24, 25; "light organ" in, 86

Marine Biological Laboratory (Woods

Hole): animal-bacteria symbiosis lecture, 84–85; author at, 11; cephalopod camouflage research, 92–93; fetal development research, 108; neuroscience research, 98–100, 107, 109, 111, 116–17, 125–29

Marshall, Greg, 68

Martha's Vineyard, 111

mating and reproduction, 23, 25, 132–46; *133*; animal-bacteria symbiosis, 85, 87; and bioluminescence, 79; and brooding, 25; death, after, 140–41, 165, 169, 174; guard males, 137, 190; hypodermic reproduction, 139; and mating, 136–40; sneaker males, 137–38

maze, cuttlefish, 191–92, 194; *191*

McFall-Ngai, Margaret, 84–87, 176

McQuhae, Peter, 16–17

medical research: on addiction, 25; Alzheimer's, 11, 81, 111, 129, 130; animal-bacteria symbiosis, 84–85; bowel disease, 88; cancer treatments, 82–83, 92, 108–9; chromosome-based diseases, 121; human fetal research, 87–88, 103, 108; and neuroscience, 81, 104–5, 112–13, 124–31. *See also* medicines

medicines, 53–54, 80–83; antibiotics, 88, 121–22; for breast cancer, 83; channel blockers, 119; penicillin, 121–22, 126; tranquilizers, 119

Menard, Wilmon, 153

Menashi, Wilson P., 147–48, 150, 155–58, 176; *148*

mercury levels, 57

Mesonychoteuthis hamiltoni (colossal squid), 22, 27, 51, 72, 83, 101, 152, 170

microscopy, video-enhanced, 125

migration, water-column, 28, 144

migration patterns, changes in, 144–45, 196–98

military research: camouflage, 91–93

Millersville University (Pennsylvania), 190

mollusks (*Mollusca*), 19–21, 23; *Aplysia*,

ABOUT THE AUTHOR

WENDY WILLIAMS is the author of several books, including the recent *Cape Wind: Money, Celebrity, Class, Politics, and the Battle for Our Energy Future on Nantucket Sound.* Her journalism has appeared in *Scientific American, Science,* the *Wall Street Journal,* the *New York Times, Parade* magazine, *Conservation Biology,* the *Boston Globe,* and in many other publications. She has won a number of awards for investigative reporting, and in 2007 *Cape Wind* was named one of the year's ten best environmental books by *Booklist* and one of the year's best science books by *Library Journal.* She lives in Mashpee, Massachusetts, on Cape Cod.

EDITOR: David Cashion
DESIGNER: Sarah Gifford
PRODUCTION MANAGER: Alison Gervais

Library of Congress Control Number: 2010032489

Paperback ISBN: 978-0-8109-8466-0
eISBN: 978-1-61312-085-9

Printed and bound in the United States
10 9 8 7 6 5 4 3 2 1

ABRAMS The Art of Books
195 Broadway, New York, NY 10007
abramsbooks.com